**Handbuch der Zoologie**
**Handbook of Zoology**
Band/Volume IV Arthropoda: Insecta

Timothy R. New/Günther Theischinger
Megaloptera (Alderflies, Dobsonflies)
Teilband/Part 33

# Handbuch der Zoologie

Eine Naturgeschichte der Stämme des Tierreiches

# Handbook of Zoology

A Natural History of the Phyla of the Animal Kingdom

Gegründet von / Founded by Willy Kükenthal
Fortgeführt von / Continued by M. Beier, M. Fischer, J.-G. Helmcke,
D. Starck, H. Wermuth

---

**Band/Volume IV Arthropoda: Insecta Teilband/Part 33**

Herausgeber/Editor Maximilian Fischer

W DE G Walter de Gruyter · Berlin · New York 1993

Timothy R. New
Günther Theischinger

# Megaloptera
## (Alderflies, Dobsonflies)

Walter de Gruyter · Berlin · New York 1993

*Autoren/Authors*

Dr. Timothy R. New
Department of Zoology
La Trobe University
Bundoora, Victoria 3083
Australia

Dr. Günther Theischinger
20 Leawarra Street
Engadine, New South Wales 2233
Australia

*Herausgeber und Schriftleiter*
*Scientific and Managing Editor*

Hofrat Univ.-Doz.
Mag. Dr. Maximilian Fischer
Direktor am
Naturhistorischen Museum Wien
Burgring 7
A-1014 Wien
Austria
Tel. (0222) 52177-316

*Verlag/Publishers*

| | |
|---|---|
| Walter de Gruyter & Co. | Walter de Gruyter, Inc. |
| Genthiner Straße 13 | 200 Saw Mill River Road |
| D-1000 Berlin 30 | Hawthorne, N.Y. 10532 |
| Tel. (030)26005-0 | USA Tel. (914)747-0110 |
| Telefax (030)26005-251 | Telefax (914)747-1326 |

Dieser Band enthält 98 Abbildungen.
This volume contains 98 figures.

♾ Printed on acid-free paper which falls within the guidelines of the ANSI to ensure permanence and durability.

*Die Deutsche Bibliothek* – Cataloging in Publication Data

**Handbuch der Zoologie** : eine Naturgeschichte der Stämme des Tierreiches / gegr. von Willy Kükenthal. Fortgef. von M. Beier ... – Berlin ; New York : de Gruyter.
Teilw. mit Parallelt.: Handbook of zoology
NE: Kükenthal, Willy [Begr.]; Beier, Max [Hrsg.]; PT

Bd. 4. Arthropoda: Insecta / Hrsg. Maximilian Fischer.
Teilbd. 33. Megaloptera : (Alderflies, Dobsonflies) / Timothy R. New ; Günther Theischinger. – 1993
ISBN 3-11-013566-3
NE: Fischer, Maximilian [Hrsg.]; New, Timothy R.

# Contents

# Megaloptera (Alderflies, Dobsonflies)

## Historical account

The Megaloptera have long been recognised as a distinct segregate of neuropteroid insects, but their formal status has varied between that of suborder and order. They are clearly related to both Raphidioptera and Planipennia and most workers today opt to recognise each of these three groups as having full ordinal status. The category Megaloptera was first recognised by ***Latreille***, who also established the genera typifying the three major groups as they are recognised today, namely *Chauliodes, Corydalus* and *Sialis* (***Latreille*** 1796, 1802, 1803). Species now included in all three genera were described earlier, under *Hemerobius* (***Linnaeus*** 1758, 1763) (the taxa now known as *Chauliodes pectinicornis, Corydalus cornutus* and *Sialis lutaria*), and the first Oriental species was described soon after this (*Protohermes grandis*: ***Thunberg*** 1781). ***Latreille***'s 'Megaloptera' also included *Raphidia* and was regarded as a family of the Planipennia.

Four groups included under Planipennia by ***Burmeister*** (1839) were raised by ***Handlirsch*** (1908) to ordinal rank – so that Sialidae became Megaloptera, Megaloptera became Neuroptera, Raphidioidea changed status, and Panorpinae became Mecoptera. In spite of various attempts to reunite Megaloptera (under various names) and Raphidioptera in an enlarged concept of Megaloptera (as examples, by ***van der Weele*** 1910, ***Tillyard*** 1918, ***Withycombe*** 1925, ***Eglin*** 1940), ***Handlirsch***'s concept of three orders within the superorder Neuropteroidea (or Neuropterida: ***Kristensen*** 1981) is now widely accepted. In this case, the name 'Planipennia' is rightly replaced by 'Neuroptera', as emphasised by ***Mansell & Aspöck*** (1990). Megaloptera are treated here as discrete order.

As with other neuropteroids, many of the genera and species of Megaloptera were described during the nineteenth century, commonly in papers including other groups of insects as well and by workers whose scope encompassed many of these under the general Linnaean scope of 'Neuroptera'. Many of these descriptions, although brief and based on a limited range of structural characters, have stood the test of time well. Most authors (including ***P.A. Latreille, G.R. Gray, F.-J. Pictet, E. Newman, H.H.C. Burmeister, J.P. Rambur, W.F. Erichson, J.O. Westwood, H.A. Hagen, R. McLachlan, F. Brauer, H.J. Kolbe*** and ***J. Wood-Mason***) were based in Europe at the time they described new species or genera, but a few (***N. Banks, T. Say***) were in North America. Together, they laid a solid framework for taxonomy of Megaloptera from many parts of the world before the beginning of the twentieth century, and facilitated later more comprehensive appraisals. ***McLachlan*** (1869) enumerated 2 genera and 37 species of Corydalidae.

Descriptive work continued to dominate work on Megaloptera for the first few decades of this century. From about 1930 onwards, although few workers have studied systematics of Megaloptera at any one time, more comprehensive revisionary or regional studies emerged. The pioneer of this class of contribution is the Monographic Revision of Megaloptera (including Raphidioptera) by ***H.W. van der Weele*** (1910), an invaluable survey of all earlier taxonomic work which remains indispensable to present-day researchers. ***Van der Weele*** included the insects currently placed in Megaloptera in the Sialidae (following ***K.C. Davis***, 1903, as Sialididae) and included Raphidioptera as the other family (Raphidiidae). He presented keys to genera, and gave detailed appraisals, with genitalic information and illustrations where available, of 70 species in 18 genera. The four photographic plates in that Monograph together show well the appearance of 40 taxa of Megaloptera: ***van der Weele***'s treatment of Raphidiidae was deliberately brief following, as it does, the more comprehensive appraisal of these by ***Albarda*** (1891). Amongst others who contributed to knowledge of the Megaloptera in the early twentieth century were ***N. Banks, G. Enderlein, H. Stitz*** and ***L. Navás*** (notorious amongst students of Neuroptera *sensu* ***Linnaeus*** for his brief descriptions of numerous taxa: he described 3 genera, 50 species and 6 'varieties' of Megaloptera [***Monserrat*** 1986]).

A number of workers have augmented taxonomic perspective more recently: they include ***Ross*** (1937), ***Hazard*** (1960), ***Flint*** (1965), ***Glorioso*** (1981), ***Glorioso & Flint*** (1984) and ***Penny & Flint*** (1982). ***Kimmins*** (1954) produced a key to the genera of Corydalidae: Chauliodinae (as Chauliodini) and ***Glorioso*** (1981) to the Corydalidae: Corydalinae, together with a generic level cladistic appraisal of this group.

Important faunistic studies and appraisals include accounts for Europe (***Aspöck*** et al. 1980, as part of their monumental *Die Neuropteren Europas*), eastern Palaearctic (***Vshivkova*** 1979–1985), Japan (***Kuwayama*** 1962, 1964, 1965), Indochina (***Lestage*** 1927), China (***Yang & Yang*** 1986, 1988, 1989, 1990), North Africa (***Aspöck & Aspöck*** 1983), South Africa (***Barnard*** 1931, 1940, ***Tjeder*** 1974), North and South America (***Davis*** 1903, ***Ross*** 1937), the Caribbean and/or parts of South America (***Flint*** 1970, 1973, 1983, ***Geijskes*** 1984, ***Penny*** 1977, 1981, 1982) and Australia (***Tillyard*** 1918, ***Riek*** 1954, ***Theischinger*** 1983, ***Theischinger & Houston*** 1988).

Most of the abovementioned accounts are based entirely or largely on adults, but immature stages of the three major divisions of Megaloptera were also recognised early on (***Riley*** 1876, 1878, 1879, ***Moody*** 1878) and have since been described in much greater detail (p. 61). They emphasise the evolutionary interest of Megaloptera as one of the most primitive kinds of 'basic' endopterygote insects.

Early biological studies on Megaloptera were almost wholly on species living in Europe or North America, as were some detailed studies on aspects of morphology, anatomy, embryology and physiology. These will be referred to specifically in later sections of this account, but comprehensive appraisals of all then known of the order were given by ***Stitz*** (1914), ***Handlirsch & Beier*** (1936) and ***Friedrich*** (1953), as well as in a number of more general texts on entomology and freshwater biology. Thus, although there is a strong regional bias in knowledge of many aspects of the Megaloptera, the basic framework of available information is strong, and they can be considered relatively well-known in comparison with some other 'small orders' of insects. The general and regional appraisals noted above provide very useful foundations for more detailed biological appraisals, and many of the taxa described can be recognised with reasonable confidence.

## Diagnosis of order

Moderate-sized to very large insects (forewing length ca 8–85 mm), comprising about 300 recent species in 2 families. A primitive group of endopterygote insects with aquatic larvae, exarate decticous pupae, and adults usually occurring close to water. Oviparous, bisexual.

## Adults

Head prognathous, generally broad and flat; large laterally projecting compound eyes; large unarmed or expanded and armed postocular region; 3 ocelli present in Corydalidae, absent in Sialidae; chewing mouthparts, unspecialised except for considerable enlargement of mandibles in males of some Corydalidae; antennae many-segmented filiform, moniliform, serrate or pectinate (commonly sexual dimorphism in antennal and mandibular form). Pronotum large, subrectangular or trapezoidal; pterothorax with meso- and metathorax homonomous. Legs cursorial, unmodified; five tarsal segments, the fourth tarsal segment of Sialidae laterally and ventrally expanded; paired apical claws. Two pairs of wings; venation complete, with little 'end-twigging'; pterostigma absent or only weakly defined; hindwing with large plicate anal field. Abdomen soft, flexible, the anterior 8 segments similar in form; tergite IX of Sialidae and Corydalidae: Corydalinae with gonopods, of Corydalidae: Chauliodinae without gonopods; tergite IX of females with gonopods forming an ovipositor; tergite X of both sexes forming a pair of anal plates. Representative adults are shown in Figs 1–3.

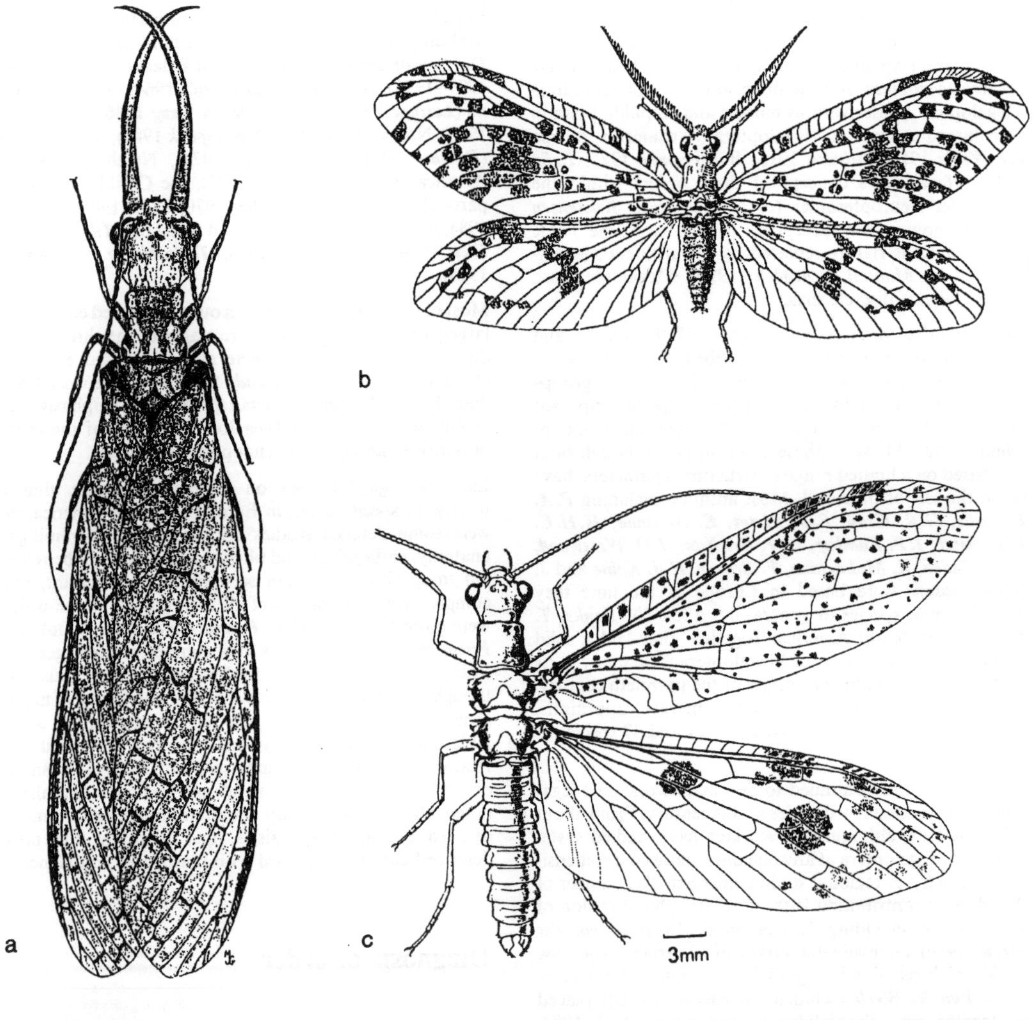


Fig. 1. Adult Corydalidae: a, *Corydalus cornutus*, male; b, *Neochauliodes sinensis;* c, *Archichauliodes* sp. (a, **Glorioso** 1981; b, **Berland & Grassé** 1951; c, **Riek** 1970, **Theischinger** 1991).

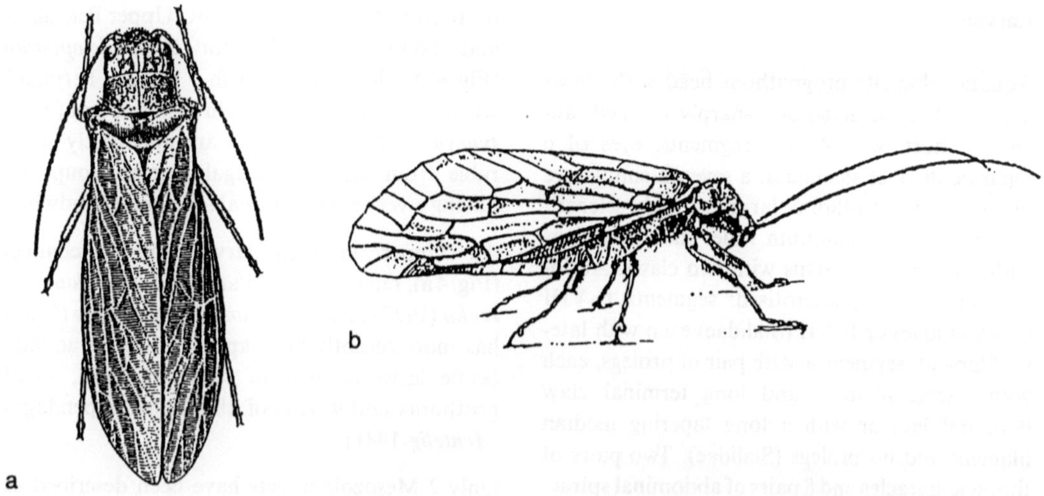

Fig. 2. *Sialis*: a, dorsal; b, lateral (a, ***Davis*** 1903; b, ***Berland & Grassé*** 1951).

Fig. 3. General appearance of selected Corydalidae. A, *Corydalus*; B, *Archichauliodes*; (forewing length: A, 45 mm; B, 32 mm).

### Larvae

Aquatic; elongate prognathous head with chewing mouthparts, mandibles sharply toothed; antennae short, with 4 or 5 segments; eyes of 6 separate distinct stemmata, a seventh simple eye often present. Prothorax large, strongly sclerotised; meso-and metanotum wider than long; legs with unsegmented tarsus with two claws. Abdomen more lightly sclerotised: segments I–VIII (Corydalidae) or I–VII (Sialidae) each with lateral filament; segment X with pair of prolegs, each with lateral filament and long terminal claw (Corydalidae) or with a long tapering median filament and no prolegs (Sialidae). Two pairs of thoracic spiracles and 8 pairs of abdominal spiracles. Representative larvae are shown in Figs 68, 69 (p. 61, 62).

## Palaeontology and phylogeny

Megaloptera evidently have a long evolutionary history although, in general and in common with many other 'neuropteroid' fossils, many of the early remains cannot be appraised with great confidence and their identities and affinities are uncertain. Many of the fossils initially designated as Megaloptera have later been transferred to other insect orders.

The Upper Permian species of *Permosialis*, described from Europe (***Martynov*** 1928), Siberia (***Martynova*** 1952) and Australia (***Riek*** 1968), have been considered to represent the oldest Megaloptera (***Rohdendorf*** 1969). ***Carpenter*** (1954) had earlier expressed doubts over the position of the Permosialidae, as did ***Riek*** (1968), and the latter transferred this family (with the three constituent genera *Permosialis, Permonikia* and *Permonka*) to the Miomoptera (***Riek*** 1976). This step was supported by ***Rasnitsyn*** (1977), and ***Hennig*** (1981) pointed out that *Permosialis* wings (Fig. 4b) have vein Sc widely separated from R1 even at the costa whereas in all recent Megaloptera the distal part of Sc is fused with R1 for a short distance: there is no conclusive evidence for *Permosialis* being a true megalopteron. Other taxa remain anomalous (see ***Zalessky*** 1926).

***Riek***'s (1970) tentative transfer of two families formerly placed in the Mecoptera (Tomiochoristidae, Choristosialidae) to Megaloptera remains unsupported. More recently, ***Ponomarenko*** (1977) erected two Permian families which he placed in the Megaloptera. Parasialidae was established for *Parasialis* (Fig. 4d, e) and *Sojanasialis* (Fig. 4g) (both from the northern USSR, Upper Permian), and Tychtodelopteridae for *Tychtodelopterum* (Fig. 4c) (Kuzneksk Basin, Upper Permian) which was formerly placed in Miomoptera (***Martynova*** 1961). Parasialidae are apparently separable from all other megalopteran groups by having a very extensive anal lobe in the hindwing.

An Upper Permian larva from Archangel (Fig. 4a), referred to the Parasialidae by ***Ponomarenko*** (1977) and to *Permosialis* by ***Sharov*** (1953) has more recently been appraised as an aquatic beetle larva because of its small head, small prothorax and 9 pairs of abdominal appendages (***Achtelig*** 1981).

Only 2 Mesozoic insects have been described as larvae of Megaloptera. One, in poor condition, from the Triassic of North America (*Mormolucoides*) (***Hitchcock*** 1858), and the other from the early Cretaceous of China (*Coptoclava*) (***Ping***, 1928) and later also recognised as a beetle larva (***Ponomarenko*** 1961).

***Riek*** (1974) recorded the first wing of a Mesozoic megalopteron from the Upper Triassic of South Africa, and established a new family (Euchauliodidae) for this, *Euchauliodes distinctus* (Fig. 5). It differs from Recent Megaloptera in the more basal origins of veins M and Cu and the greater number of crossveins. ***Riek*** (1974) also established the suborder Eumegaloptera, but without discussing the classification of remaining Megaloptera.

Both adults and the larva of *Cretochaulus lacustris* (Figs 8, 9) were described from the Lower Cretaceous of Transbaikalia (***Ponomarenko*** 1976). This taxon belongs to the Corydalidae. ***Ponomarenko*** also described *Chauliosialis sukatshevae* (Fig. 10) from fossil resin of the Upper Cretaceous of the Taymyr National Territory. This taxon, although resembling Sialidae in the retention of abdominal segment X, had only 6 pairs of abdominal appendages and, from the presence of pygopods and its supposed respiratory mechanism, was also regarded as a corydalid. It thus differs markedly from all known Recent Megaloptera.

Very few Tertiary megalopteran fossils have been recorded. Both Recent families are present in Baltic amber, but represented by few specimens (***Larsson*** 1978). One such specimen (Oligocene) was described as *Chauliodes prisca* (***Pictet*** 1856, see also ***Hagen*** 1854, ***Riek*** 1970) and another (Upper Eocene) is a *Sialis* larva (***Weidner*** 1958). Species described in the genus *Semblis* (= *Sialis*) by ***Burmeister*** (1832) and ***Gravenhorst*** (1834) are probably Planipennia (***Hagen*** 1856).

Eggs larger than those of *Corydalus*, found in

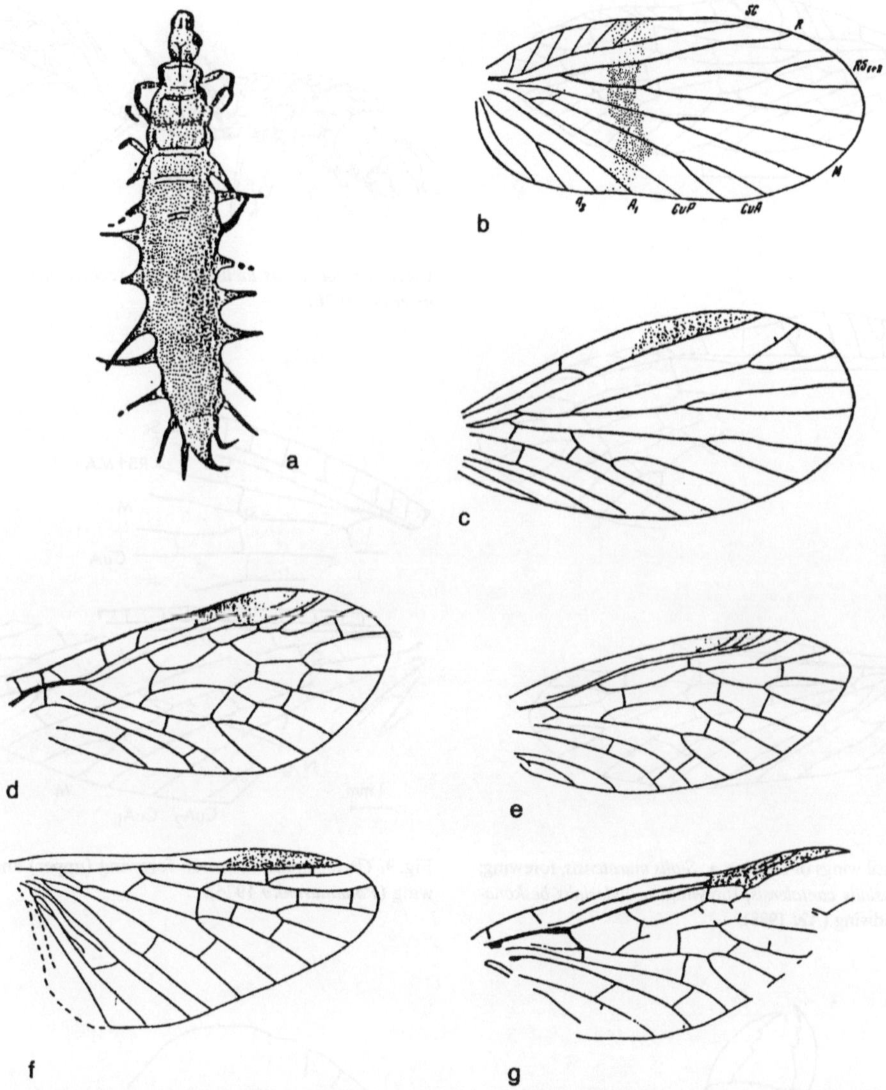


Fig. 4. Some fossils which have been referred to Megaloptera: a, possible Permian larva; b, *Permosialis fasciata*, forewing; c, *Tychtodelopterum archaicum*, forewing; d, *Parasialis latipennis*, forewing; e, *Parasialis dissendens*, forewing; f, Parasialidae indet. genus & sp.; g, *Sojanasialis longipennis*, forewing (***Martynova*** 1952; ***Ponomarenko*** 1977; ***Sharov*** 1953).

lignitic beds in Colorado, were described as *Corydalites fecundum* (***Scudder*** 1878). The batches of about 2000 eggs were apparently the first fossils of insect eggs to be found. ***Illies*** (1967), described *Sialis strausi* (Fig. 7) from Late Pliocence deposits of Willershausen (Germany), where it was common and represented by 'hundreds' of larval mandibles, and ***Nel*** (1988), three species of Sialidae from Oligo-Miocene of France and Turkey (Fig. 6); one was believed to represent a new genus, *Proindosialis*.

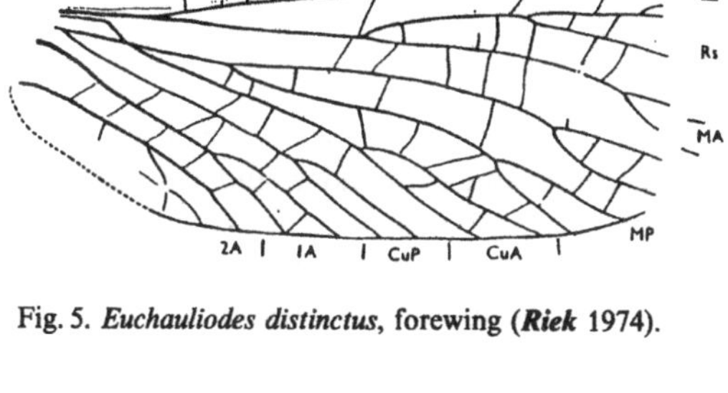


Fig. 5. *Euchauliodes distinctus*, forewing (***Riek*** 1974).

Later *Sialis* remains are widely distributed and have been recorded on many occasions – for examples from Austria (***Gams*** 1927), Germany (***Paul & Rouff*** 1927, 1932) and the United States (***Coope*** 1961). A subfossil *Sialis* larva from interglacial strata in Pilgrimstad, Sweden, (possibly about 100,000 years old) is probably a recent species (***Tjeder*** 1968).

Initial forms of Megaloptera were apparently similar to the Miomoptera with wings practically

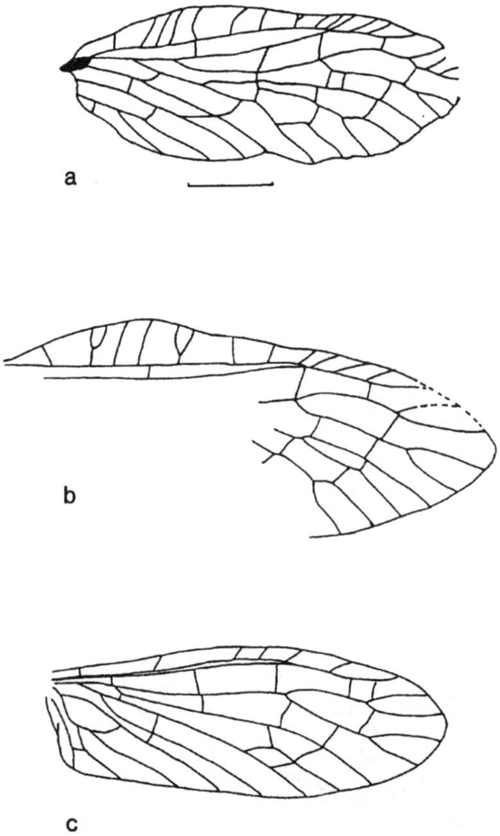

Fig. 6. Fossil wings of Sialidae: a, *Sialis muratensis*, forewing; b, *Proindosialis cantalensis*, forewing; c, *Indosialis beskonakensis*, hindwing (***Nel*** 1988).

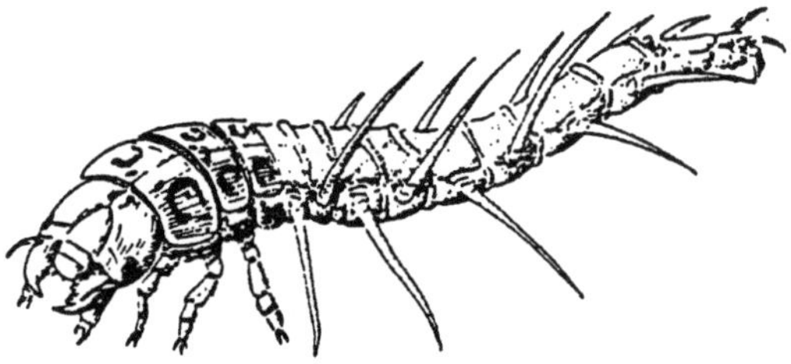

Fig. 8. *Cretochaulus lacustris* larva: reconstruction (***Ponomarenko*** 1976).

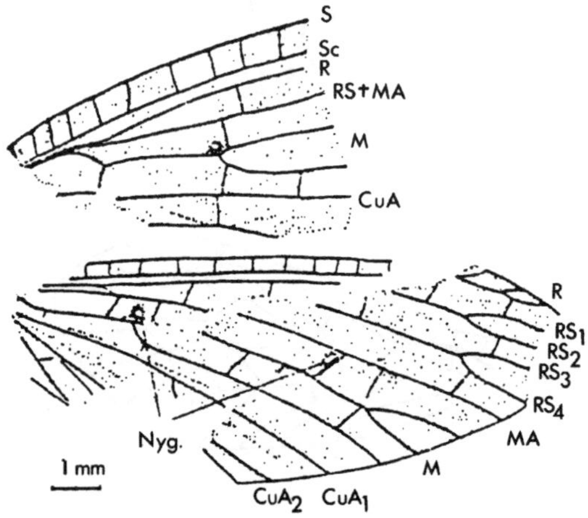


Fig. 9. *Cretochaulus lacustris*: forewing (upper) and hindwing (***Ponomarenko*** 1976).

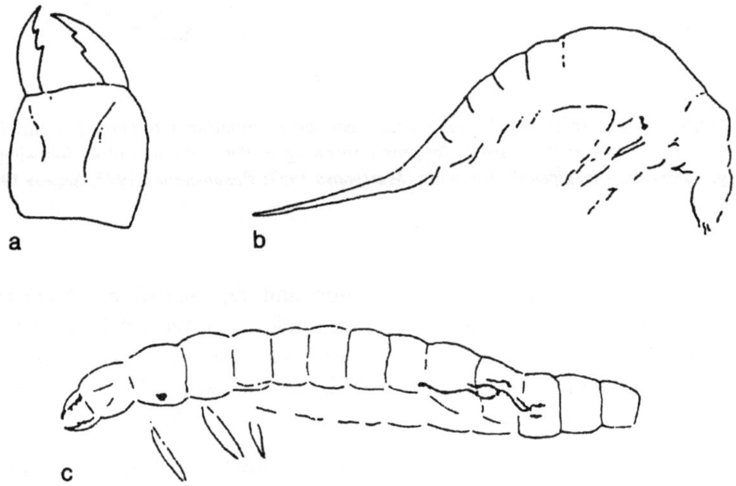

Fig. 7. *Sialis strausi* larva: a, head and mandibles; b, c, two specimens in lateral aspect (***Illies*** 1967).

homonomous, and resembled *Tychtodelopterum* except for the vein MA entering the radius more distally, MP forked only once and cubital and anal veins simple (***Ponomarenko*** 1977). The order apparently divided into two main branches at a very early stage. One is represented by the Tychtodelopteridae, Euchauliodidae and Corydalidae, and the other by Parasialidae and Sialidae. ***Ponomarenko*** (1977) suggested that Corydalidae developed at the beginning of the Mesozoic whereas the

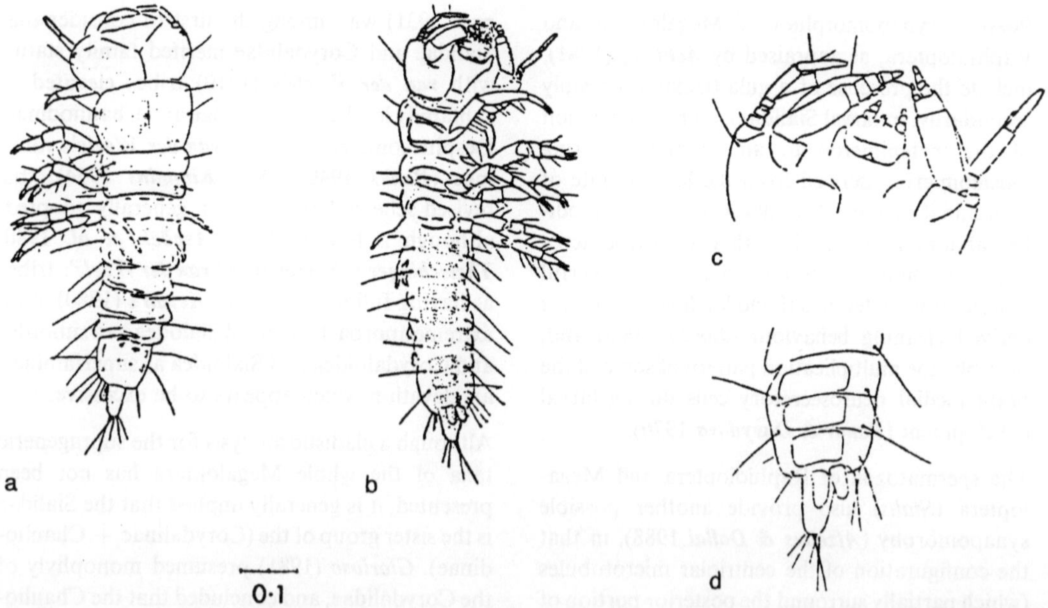


Fig. 10. ***Chauliosialis sukatshevae*** **larva: a, dorsal; b, ventral; c, head, ventral; d, apex of abdomen, ventral (*Ponomarenko* 1976).**

characteristic larvae of Sialidae developed later (possibly in the Cenozoic), and that polymerisation of venation in connection with increase in body size predominated in the evolution of Megaloptera. According to ***Ponomarenko*** (1977), the known Parasialidae cannot be regarded as ancestors of the Sialidae.

The monophyly of the three major groups of Neuropteroidea is now almost universally accepted, although synapomorphies are not prominent. These include (1) ovipositor with third valvulae fused and with a series of short intrinsic muscles; (2) metanotum divided medially; (3) abdominal tergite I with a posteriorly-bifid median longitudinal suture (***Achtelig*** 1975, 1981); (4) proventriculus with unpaired dorsal diverticulum. However, discussions of the relationships of the 3 groups here considered as orders within the Neuropteroidea (or Neuropterida) continue, especially the relative ordinal or subordinal status of Megaloptera and Raphidioptera. ***Achtelig***'s (1967) arrangement of Megaloptera recognised two suborders: Sialoidea contained only Sialidae, but Chaulioidea contained three families, Chaulioidae, Corydalidae and Raphidiidae. *Raphidia* and *Corydalus* were considered to form a natural group clearly separable from *Sialis*, and this conclusion was derived from a study of numerous head features and associated anatomy. This arrangement was revised by ***Achtelig & Kristensen*** (1973) (although acceptable to several authors during the intervening years), who recognised that a number of the shared characters could be due to convergence. A few characters were then considered evidence for a sister-group relationship between Raphidioptera and Megaloptera s.str: the presence of a gula in the adult, the telotrophic nature of the meroistic ovarioles and, possibly, cleaning behaviour. Other characters indicated a sister-group relationship between Raphidioptera and Planipennia so that ***Achtelig & Kristensen*** considered it impossible to decide which of the above sister-group pairs might be valid. Later studies on the abdominal base and female genitalia by ***Achtelig*** (1975, 1976, 1978) tend to support the closer alliance of Raphidioptera and Megaloptera as the joint sister group of Planipennia, as depicted by ***Kristensen*** (1981). However, ***Boudreaux*** (1979) had then advanced the premise that Megaloptera + Planipennia was a sister group of the Raphidioptera. The alleged synapomorphies supporting this, traits doubted by ***Kristensen*** (1981), were (1) fusion of veins Sc and R1, (2) absence of a pigmented pterostigma (3) elongate stipes in adult (not in Sialidae: ***Kristensen*** 1981) and (4) shortening of the ovipositor.

Thus, although Megaloptera and Raphidioptera have been placed as separate orders, sometimes without convincing evidence, there is now strong suggestion that they are indeed distinct. Historically, they were treated as separate by, for examples, ***Weber*** (1933), ***Handlirsch & Beier*** (1936), ***Berland & Grassé*** (1951) and ***Aspöck & Aspöck*** (1971). Others retained them together (e.g. ***Richards & Davies*** 1957, and a number of other standard entomology texts).

Possible synapomorphies of Megaloptera and Raphidioptera, as appraised by ***Achtelig*** (1981), include the presence of a gula (lacking, possibly secondarily, in larval Sialidae), a firm connection of the metepimeron to an apophysis-bearing 'postepimeron' derived from the laterotergite of abdominal segment I, a characteristically modified abdominal tergite II with a reinforced acrotergite accommodating the origin of a powerful muscle between tergites II and V (***Achtelig*** 1975), a derived cleaning behaviour (***Jander*** 1966) and, possibly, the multiplication pattern of some of the brain medial neurosecretory cells during larval development (***Panov & Davydova*** 1976).

The spermatozoa of Raphidioptera and Megaloptera (*Sialis*) also provide another possible synapomorphy (***Afzelius & Dallai*** 1988), in that the configuration of the centriolar microtubules (which partially surround the posterior portion of the nucleus: p. 42) in both differs from those representatives of Planipennia which have been examined.

Adult autapomorphies of Megaloptera remain obscure. The aquatic mode of life of the larvae and associated morphological specialisations such as slender lateral abdominal gills and the divided stipes indicate that they are likely to be a monophyletic group.

## Systematic account

***Latreille***'s (1802) establishment of three distinct genera, *Corydalus* (often, incorrectly, spelled '*Corydalis*', a name dating from ***Latreille***, 1804), *Chauliodes* and *Sialis*, first presented a reasonable reflection of modern classification of the Megaloptera; each of these now typifies a major grouping of the order recognised by many later authors. ***Davies*** (1903) divided the insects into two subfamilies, Sialidinae and Corydalinae, on the presence or absence of ocelli and differences in the elaboration of the fourth tarsal segment and of the venation in the radial sector. ***Van der Weele*** (1910) followed this subdivision of his family Sialidae but, further, divided the Corydalinae into two tribes Neuromini and Chauliodini on features of head, antennae, wing venation and male terminalia. Although the Neuromini was recognised by many authors (including ***Esben-Petersen*** 1924, ***Barnard*** 1931, ***Kimmins*** 1948, ***Kuwayama*** 1962), ***Tillyard*** (1918) and others called this group Corydalinae even before ***Lestage*** (1927) noted that any such group name should be based on *Corydalus*. ***Tillyard*** (1918) (followed by ***Lestage*** 1927, ***Barnard*** 1931) was among the first to consider that Sialidae and Corydalidae merited family status, with ***van der Weele***'s (1910) tribes elevated to subfamily level as Corydalinae and Chauliodinae. Despite some retention of ***van der Weele***'s rankings (***Banks*** 1940, 1943, ***Kimmins*** 1954), this revised scheme has now been generally adopted, although a few workers (***Aspöck et al.*** 1980, ***Theischinger*** 1983) regarded ***van der Weele***'s tribes as having full family status. ***Kevan***'s (1980) more extreme approach included Sialodea as a suborder and Corydaloidea and Sialoidea as superfamilies: an elevation which appears to be excessive.

Although a cladistic analysis for the suprageneric taxa of the whole Megaloptera has not been presented, it is generally implied that the Sialidae is the sister group of the (Corydalinae + Chauliodinae). ***Glorioso*** (1981) presumed monophyly of the Corydalidae, and concluded that the Chauliodinae are, or contain, the sister group of the Corydalinae. Cladistic analysis of the other groups is needed in order to clarify the status of a number of dubious taxa, and a number of recent taxonomic revisions (such as those by ***Flint*** 1983, ***Glorioso & Flint*** 1984, ***Theischinger*** 1983, ***Penny & Flint*** 1982) have shown well that characters of the terminalia are invaluable for specific diagnoses.

In parentheses, it should be noted that ***Navás*** (1927) established the family Brucheiseridae in the Megaloptera. The family, overlooked by students of Megaloptera for many years, was transferred to Planipennia by ***Riek*** (1975) and the two species appear to be aberrant Coniopterygidae: they were considered in that family by ***New*** (1989).

### Key to families and subfamilies

At present, only two families of extant Megaloptera are recognised. The larger of these is separable into two well-defined subfamilies, which are occasionally cited as separate families.

1 Adults with ocelli; fourth tarsal segment simple; forewing longer than 15 mm. Larvae with 8 pairs of lateral abdominal gills; apical segment of abdomen with pair of prolegs but lacking terminal filament .... Corydalidae 2

– Adults without ocelli; fourth tarsal segment bilobed; forewing 15 mm or less. Larvae with 7 pairs of lateral abdominal gills; apical segment of abdomen without prolegs but with a terminal filament .................... Sialidae

2 Adults with head subquadrate with postocular ridge, spine and plane; males with well-developed ninth gonostyli. Larvae with ventral gill tufts .................... Corydalinae

– Adults with head subtriangular without postocular ridge, spine and plane; males without ninth gonostyli. Larvae without ventral gill tufts . . . . . . . . . . . . . . . . . . . . . . Chauliodinae

Corydalidae have the common names of Dobsonflies, Fishflies and (especially for larvae of *Corydalus*) Hellgrammites. Sialidae are most popularly known as Alderflies, but are also termed Ordflies or Humpbacks. Figures 11–17, as well as some figures in the Morphology section, should be used with the following keys to genera.

## Family Corydalidae

### Subfamily Corydalinae

About 100 nominal species distributed amongst 9 genera occurring in North and South America, southern Africa or the Oriental Region. Most genera need revisionary study at the species level but the genera themselves are clearly defined (**Glorioso** 1981). Some of the nominal species may well prove to be synonyms, and **Glorioso** commented that the high intraspecific variability could mean that the total number of species may be less than 50. Numbers of species cited below should not, thus, be regarded as definitive.

*Key to genera*

1 Three crossveins between veins R1 and Rs (Fig. 11) . . . . . . . . . . . . . . . . . . . . . . . . . . . 2
– Four or more crossveins between R1 and Rs (Fig. 11) . . . . . . . . . . . . . . . . . . . . . . . . . . . 5
2 Vein M3 + 4 forked (Fig. 11); postocular region explanate (Fig. 14) . . . . . . . . . . . . . . . . . . . . . . . . . . . . . . . . . . . . . . . . . *Platyneuromus*
(synonym: *Doeringia*; ca 4 species: Mexico to Panama)
– Vein M3 + 4 simple (Fig. 11); postocular region not explanate . . . . . . . . . . . . . . . . . . . 3
3 Last branch of vein Rs branched (Fig. 12) . . . . . . . . . . . . . . . . . . . . . . *Chloronia*
(10 species: Mexico to southern Brazil)
– Last branch of vein Rs simple (Fig. 12) . . . . . . . . . . . . . . . . . . . . . . . . . . . . . . . . . . . . . . . . 4
4 Male ninth gonostyles clavate (Fig. 15); male mandibles usually enlarged . . . . . . *Corydalus*
(35 described species, many names are probably synonyms: southern Canada to northern Argentina)
– Male ninth gonostyles claw-like . . . . . . . . . . . . . . . . . . . . . . . . . . . . . . . . . . . *Chloroniella*
(1 species: South Africa)
5 Occipital disc with 2 prominent spines (Fig. 13); male mandibles enlarged . . . . . . . . . . . . . . . . . . . . . . . . . . . . . . . . *Acanthacorydalis*
(ca 7 species – possibly fewer valid: Oriental Region)
– Occipital disc without spines; male mandibles never enlarged . . . . . . . . . . . . . . . . . . . . . . . 6
6 Posterior tentorial pits arcuate (Fig. 13G); postocular spine well developed; vein 1A with 2 branches (Fig. 11) . . . . . . . . . . . . . . . . . . . 7
– Posterior tentorial pits linear (Fig. 13H); postocular spine poorly developed; vein 1A with 3 branches (Fig. 11) . . . . . . . . . . . . . . . 8
7 Male ninth sternite attenuate; clypeal margin with a few setae [mostly large dark species] . . . . . . . . . . . . . . . . . . . . . . *Neoneuromus*
(ca 8 species: northwest India to West Malaysia and China)
– Male ninth sternite quadrate (Fig. 15); clypeal margin fimbriate [moderate sized pale species] . . . . . . . . . . . . . . . . . . . . . . . . *Neuromus*
(4 species: northwest India to Indonesia, China)
8 Male tenth tergites always biramous, ventral arm slightly longer than dorsal arm [head and body black, pronotum of some species orange, wings black with light maculations] . . . . . . . . . . . . . . . . . . . . . . . . . . . . . . *Neurhermes*
(synonym: *Hermes*; 8 species: northwest India to Indonesia, China)
– Male tenth tergites simple (Fig. 15) or with dorsal arm much longer than ventral arm [body usually pale, wings hyaline to fumose, sometimes with lighter markings] . . . . . . . . . . . . . . . . . . . . . . . . . . . . . . *Protohermes*
(synonym: *Allohermes*; ca 42 species: northwest India to Indonesia, China and Japan)

### Subfamily Chauliodinae

About 90 species (and 9 subspecies) distributed amongst 15 genera, occurring in North America, Chile, southern Africa and Madagascar, southeast Asia, Australia or New Zealand. Some genera appear very similar, whereas some subgeneric differences (such as between *Archichauliodes* sg. *Archichauliodes* and *Apochauliodes*) and differences between species groups (Australian *A.* sg. *Archichauliodes*) are more robust. Geographical range may be relevant as a generic character. See comment under Corydalinae regarding reliability of diversity estimates.

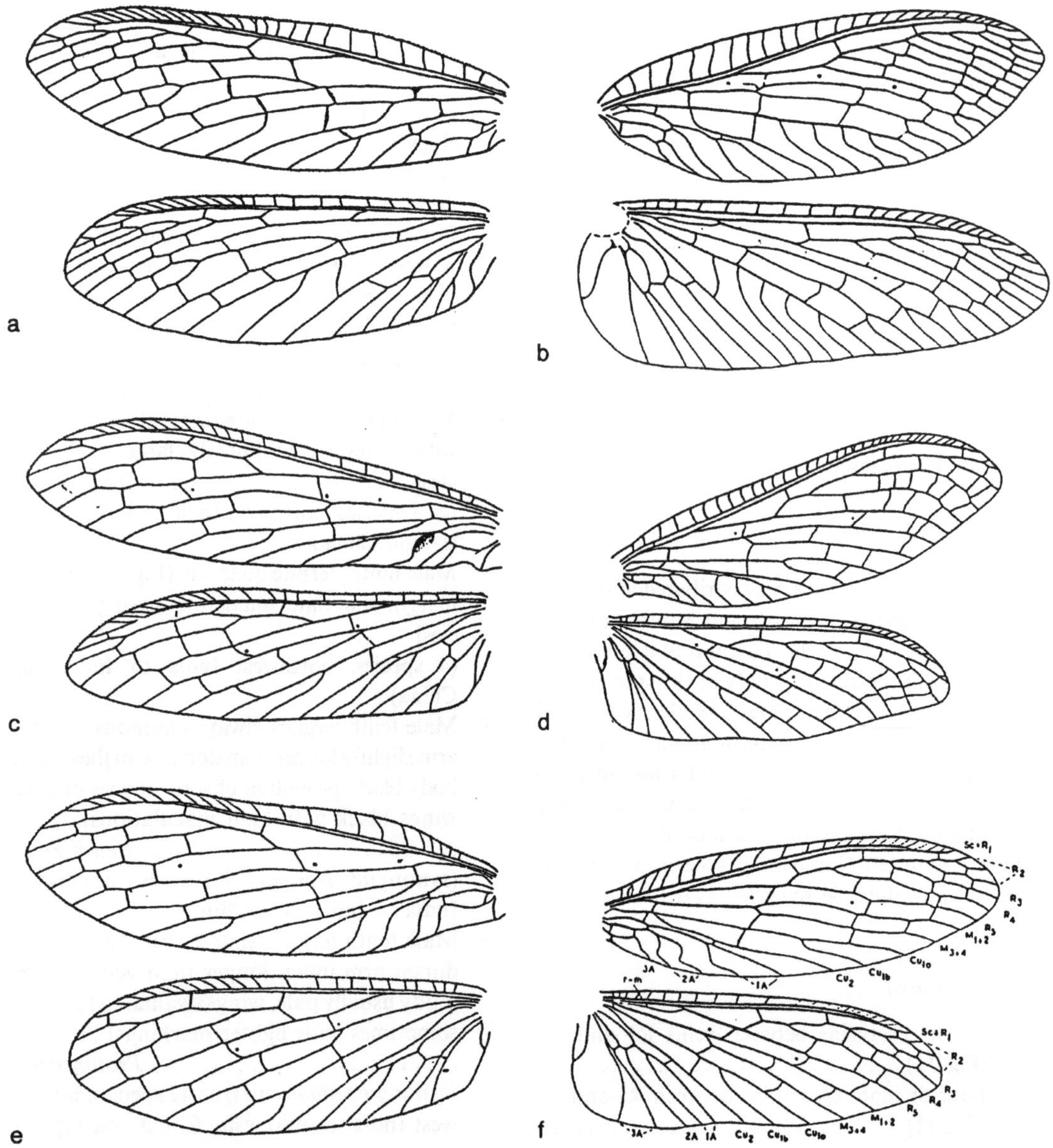


Fig. 11. Wing venation of Corydalidae: a, *Chloroniella*; b, *Platyneuromus*; c, *Taeniochauliodes*; d, *Anachauliodes*; e, *Platychauliodes*; f, *Archichauliodes* (a, c, e, **Barnard** 1931; b, **Glorioso & Flint** 1984; d, f, **Kimmins** 1954).

*Key to genera*

1 Posterior branch of vein Rs in both wings forked ............................. 2
– Posterior branch of vein Rs in both wings simple ............................. 3
2 Vein M in hindwing with 4 branches ..... ........................ *Dysmicohermes* (2 species: western USA and Canada)
– Vein M in hindwing with 3 branches (Fig. 12) .................... *Orohermes* (1 species: western USA)
3 Anterior branch of forewing vein 2A partly fused with vein 1A, so that this branch apparently arises from 1A ............. 4
– Anterior branch of forewing vein 2A not fused with vein 1A: *either* it *or* the stem of vein 2A connected to 1A by a crossvein .. ..................................... 7
4 Anterior branch of hindwing vein M simple ................ *Taeniochauliodes* (synonym: *Leptochauliodes*; 1 species: South Africa)
– Anterior branch of hindwing vein M forked ............................ 5
5 Vein R4 in both wings simple (Fig. 12) ... ....................... *Nothochauliodes* (1 species: Chile)
– Vein R4 in both wings forked .......... 6

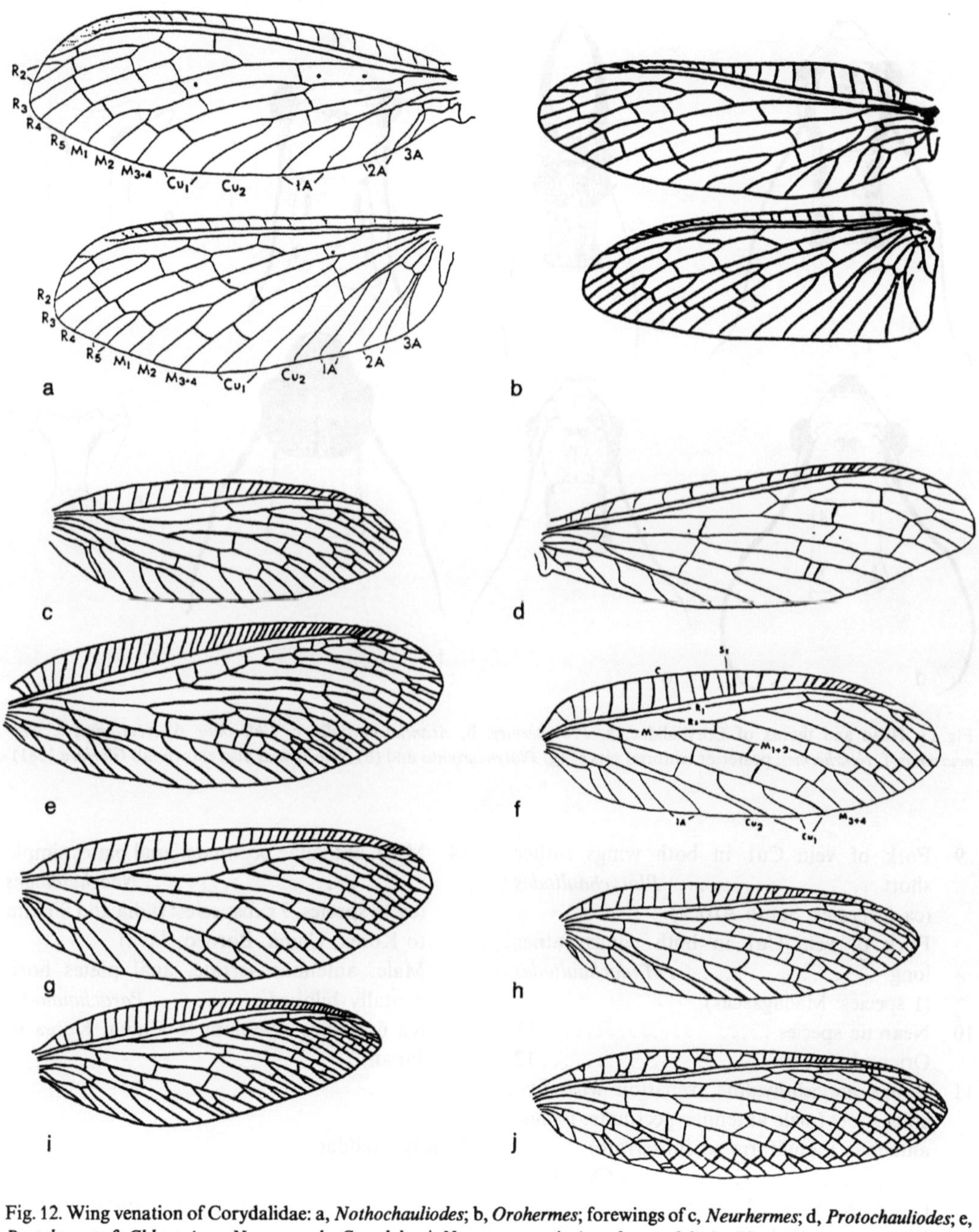


Fig. 12. Wing venation of Corydalidae: a, *Nothochauliodes*; b, *Orohermes*; forewings of c, *Neurhermes*; d, *Protochauliodes*; e, *Protohermes*; f, *Chloronia*; g, *Neuromus*; h, *Corydalus*; i, *Neoneuromus*; j, *Acanthacorydalis* (a, **Flint** 1983; b, **Chandler** 1954; d, **Kimmins** 1954; rest, **Glorioso** 1981).

6 Male: antennae with a whorl of erect hairs on segments except scape; female: lateral gonapophyses without small apical cercus .. .............................. *Neohermes*
(5 species: USA and Canada)

– Male: antennae without a whorl of erect hairs on each segment; female: lateral gonapophyses with a small apical cercus ........ ........................ *Protochauliodes*
(synonym: *Austrochauliodes*; 14 species, 3 subspecies: western North America, Chile, Australia).

7 Species from Australia, New Zealand, Chile, South Africa or Madagascar ........... 8

– Species from Nearctic or Oriental Regions ................................... 10

8 Hindwing with basal r-m crossvein ...... ........................ *Archichauliodes*
(2 subgenera: *Archichauliodes*, 20 species, eastern Australia, New Zealand, Chile; *Apochauliodes*, 1 species, Western Australia)

– Hindwing without basal r-m crossvein (South Africa or Madagascar) ......... 9

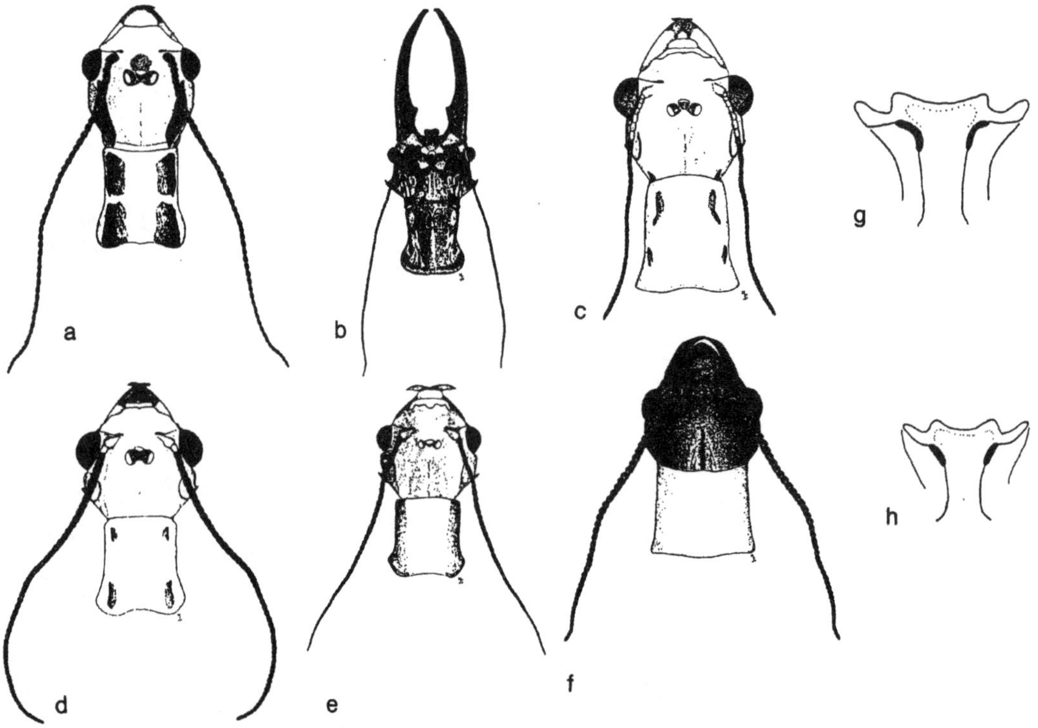


Fig. 13. Head and thorax of Corydalidae: a, *Protohermes*; b, *Acanthacorydalis*; c, *Chloronia*; d, *Neuromus*; e, *Neoneuromus*; f, *Neurhermes*; posterior tentorial pits of (g) *Platyneuromus* and (h) *Neurhermes* also shown (all **Glorioso** 1981).

9 Fork of vein Cu1 in both wings rather short .................. *Platychauliodes*
(ca 6 species; South Africa)
– Fork of vein Cu1 in both wings rather long .................. *Madachauliodes*
(1 species: Madagascar)
10 Nearctic species .................... 11
– Oriental species .................... 12
11 Wings greyish-hyaline, venation with fuscous spots (male antennae pectinate; female antennae moniliform to subserrate) ...... ........................... *Chauliodes*
(2 species: eastern USA and Canada)
– Wings mainly dark brown to black, with creamy white markings (male antennae serrate to pectinate; female antennae subserrate) ........................ *Nigronia*
(2 species: eastern USA and Canada)
12 Forewing vein 1A with 3 or 4 branches (Fig. 11) ................ *Anachauliodes*
(Male antennae pectinate, female unknown; 1 species: Vietnam)
– Forewing vein 1A with 2 branches ..... 13
13 Forewing vein 2A: fork sessile ......... ........................ *Ctenochauliodes*
(Antennae of both sexes pectinate; 6 species: China, Indochina)
– Forewing vein 2A: fork with definite basal stem ............................... 14
14 Male: antennae pectinate, anal plates simple ........................ *Neochauliodes*
(ca 18 species, 7 subspecies: India and Sikkim to Korea, China, Borneo, Java)
– Male: antennae serrate, anal plates horizontally bilobed .......... *Parachauliodes*
(ca 6 species: Vietnam to China, Korea to Japan)

## Family Sialidae

Around 60 species distributed amongst about 7 genera, from most temperate and tropical regions. Few species, mainly those in northern temperate regions, are common and tropical taxa, in particular, are rare and poorly known. The last workable key to genera was by **Lestage** (1927), and generic classification in the family is now confused and in need of a completely new analysis: at present distributional traits are as useful as many morphological features in seeking to apply generic names correctly. The following key to genera is, therefore, suggestive rather than definitive, as the validity of some of the small features used to erect genera in Sialidae is suspect. It is based on that of **Lestage** (1927): the sparse material he noted for some genera has scarcely been augmented since then.

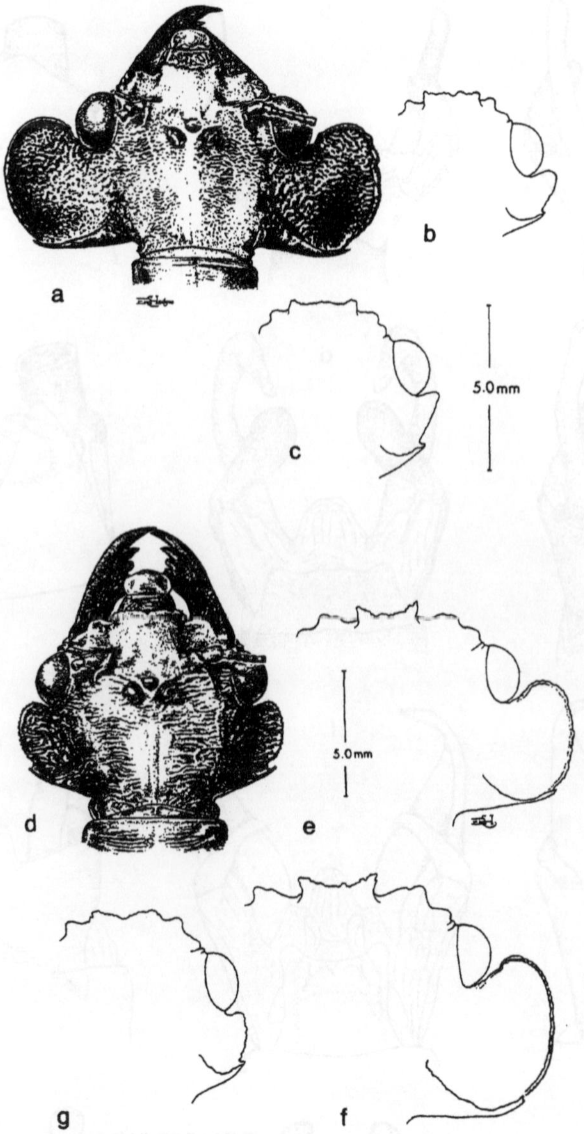


Fig. 14. Variation in head form of *Platyneuromus*: a–c, *P. honduranus*; d–g, *P. soror* (a, b, d–f, males; c, g, females) (*Glorioso & Flint* 1984).

*Key to genera*

1 Costal field dilated only at base (all veins in both wings, except R2 + 3, simple; wings long and straight; venation indistinct) (Fig. 17) .. .............................. *Leptosialis*
(1 species: South Africa)

– Costal field with more extensive dilation (venation different from above) ............ 2

2 Vein R3 forked in both wings; costal field strongly dilated (M1 + 2 simple in both wings; M3 + 4 forked in forewing, simple in hindwing) (Fig. 17) .................... *Sialis*
(Synonym: *Nipponosialis*; possibly also *Protosialis* (couplet 5) ca 52 species: Holarctic and Palaearctic)

– Vein R3 simple in both wings; costal field less dilated ............................... 3

3 Costal field with dilation clearly apparent (Forewing veins M1 + 2, M3 + 4 forked; hindwing vein M1 + 2 forked, M3 + 4 simple) ............................. 4

– Costal field with only weak dilation (Vein M1 + 2 simple in both wings) ........... 5

4 Forks of forewing veins R4 + 5, M1 + 2, M3 + 4 and CuA of similar lengths (Fig. 17) ..................... *Austrosialis*
(2 species: Australia)

– Forks of forewing veins R4 + 5, M1 + 2, M3 + 4 and CuA of markedly differing lengths ....................... *Stenosialis*
(2 species: Australia)

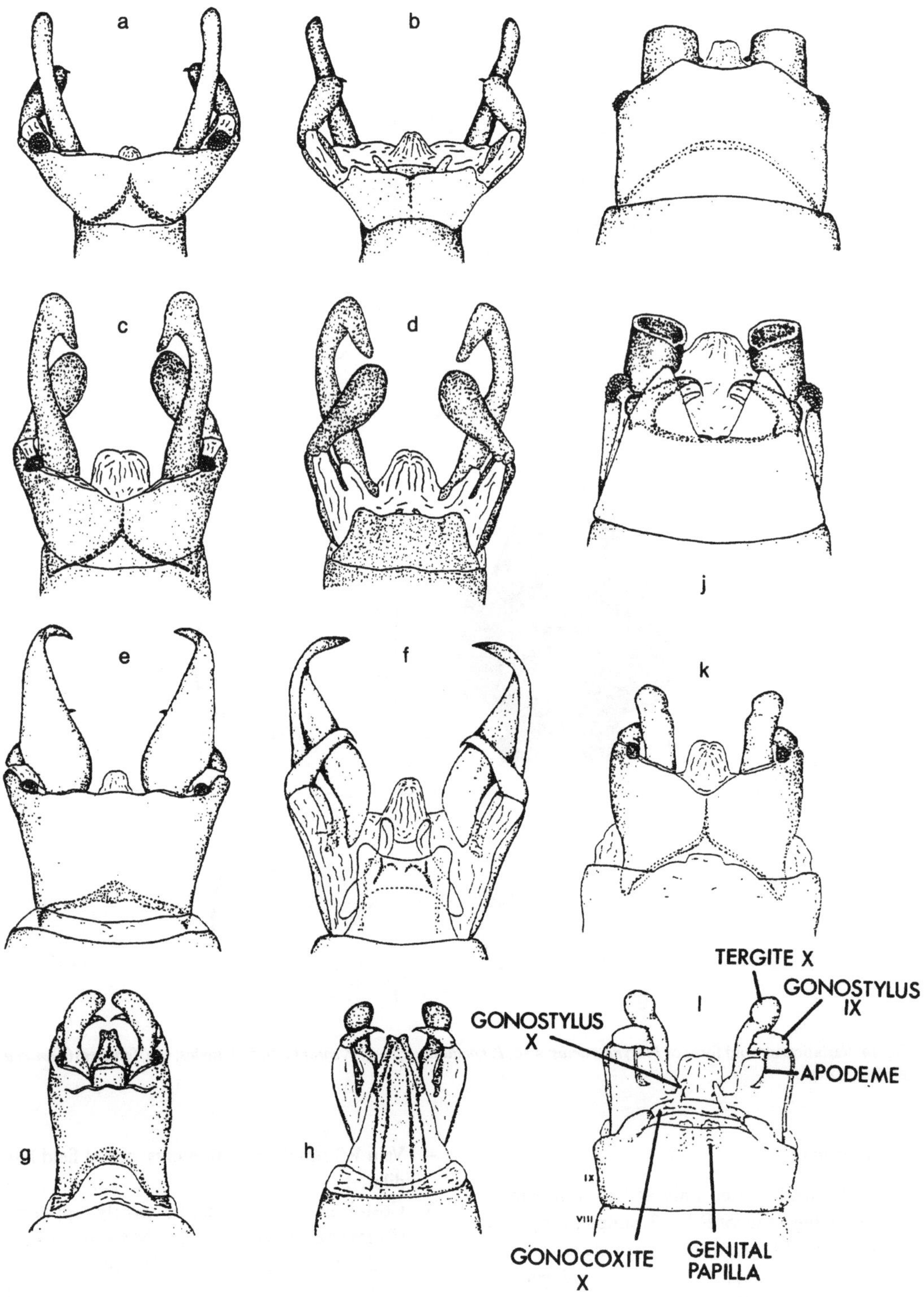


Fig. 15. Male genitalic structures of Corydalidae. a, c, e, g, i, k, dorsal; b, d, f, h, j, l, ventral: a, b, *Chloronia*; c, d, *Corydalus cornutus*; e, f, *Neuromus*; g, h, *Neoneuromus*; i, *Protohermes*; k, l, *Acanthacorydalis* (all ***Glorioso*** 1981) (see also Fig. 37).

5 Forewing vein M3 + 4 simple ............ ............................... *Indosialis*
(1 species, described as *Protosialis*: Singapore)

– Forewing vein M3 + 4 forked ............ ............................... *Protosialis*
(9 species: southern USA to Chile; synonymised with *Sialis* by ***Ross***, 1937)

*Notes:* a) *Haplosialis* is not included above. It was raised by ***Navás*** for a Madagascan species but not considered by ***Paulian*** (1951).

b) *Nipponosialis* was recognised (as '*Npposialis*') from the Soviet Far-East by ***Vshivkova*** (1979).

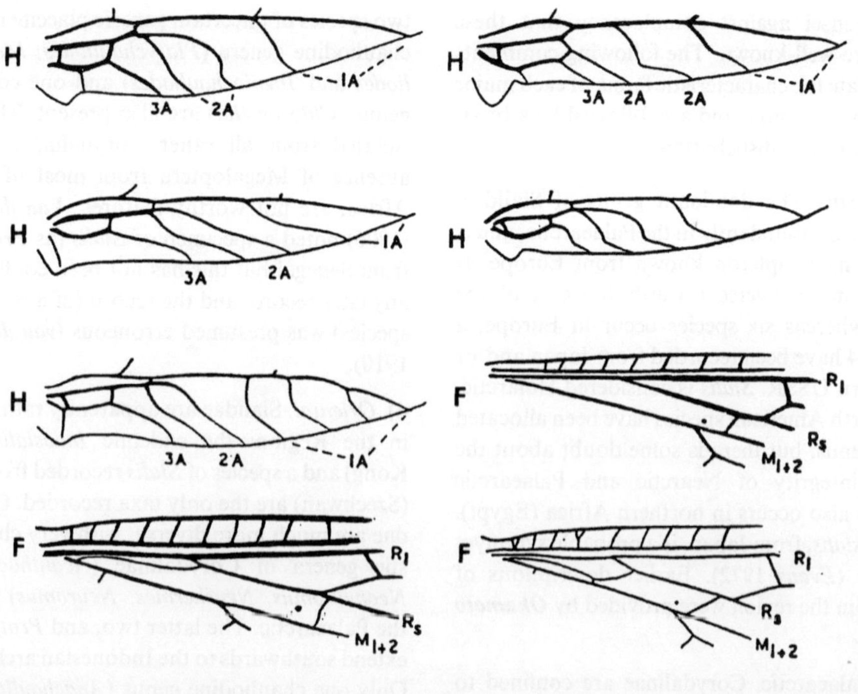


Fig. 16. Wing venation of Corydalidae: (F, forewing, anterior basal region; H, hindwing, posterior basal region of (left column, top to bottom) *Archichauliodes, Chauliodes, Anachauliodes, Dysmicohermes,* (right column, top to bottom) *Protochauliodes, Platychauliodes, Platychauliodes, Ctenochauliodes* (from **Kimmins** 1954).

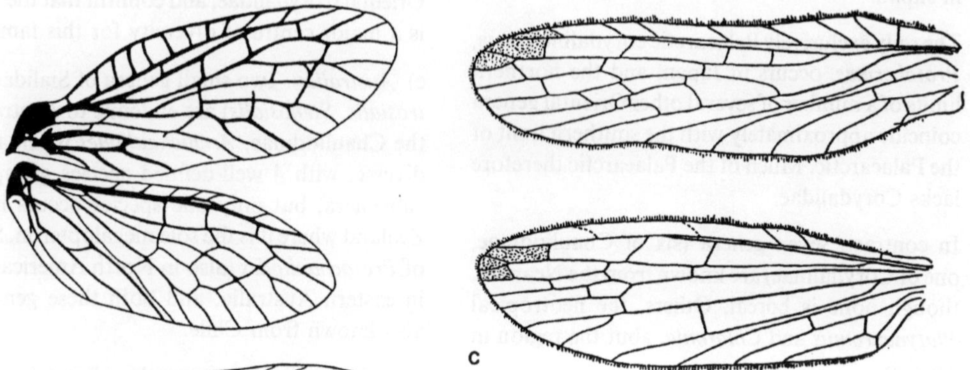

a

c

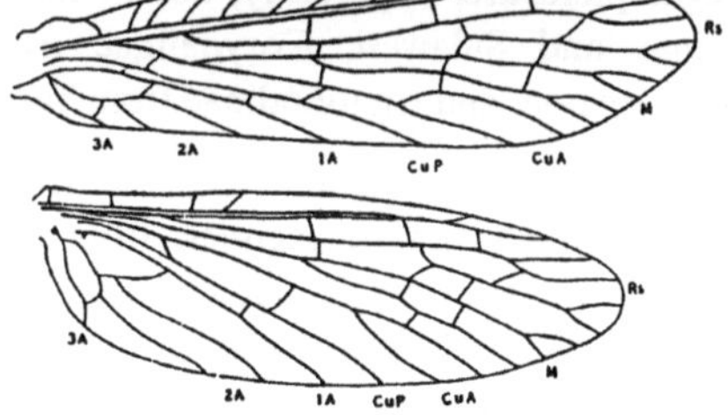


b

Fig. 17. Wing venation of Sialidae: a, *Sialis*; b, *Stenosialis*; c, *Leptosialis* (with trichiation shown in some apical cells) (a, *S. fuliginosa*, ***Berland & Grassé*** 1951; b, ***Riek*** 1970, **Theischinger** 1991; c, ***Barnard*** 1931).

## Zoogeography and distribution

Megaloptera, as a group, occur in many temperate and tropical regions of the world but – as implied in the systematic accounts – many genera and species are very circumscribed in distribution. Few genera are at all widespread and, although the doubtful integrity of some taxa, especially of Sialidae, may obscure some distributional relationships it is likely that most taxa of these often large and spectacular insects are now represented in collections. However, the recent papers by ***Yang & Yang*** (1986–1990) in which more than 20 new species are described from China and nearby areas counsel against complacency that these insects are well-known. The following comments recapitulate the characteristic fauna of each major geographical region, and are followed by a broader appraisal of distribution.

a) *Holarctic*. The dominant genus of Sialidae, *Sialis*, occurs abundantly in the Palaearctic, and is the only megalopteron known from Europe. It becomes more diverse towards the east of the region; whereas six species occur in Europe, a further 14 have been recorded from Japan and/or the eastern USSR. *Sialis* is considered Holarctic, as 20 North American species have been allocated to this genus, but there is some doubt about the generic integrity of Nearctic and Palaearctic forms. It also occurs in northern Africa (Egypt). *Nipponosialis*, from Japan, is a probable synonym of *Sialis* (***Evans*** 1972). Earlier descriptions of *Sialis* from the region were provided by ***Okamoto*** (1910).

In the Palaearctic, Corydalinae are confined to northern India, Assam, Bhutan, China, Mongolia, Korea and Japan. Members of the predominantly Oriental chauliodine genera *Ctenochauliodes, Neochauliodes* and *Parachauliodes* also occur in Japan.

The only exclusively Palaearctic corydaline genus, *Protohermes*, occurs in Japan, and the northern limits of a number of several other Oriental genera coincide approximately with the southern limit of the Palaearctic. Much of the Palaearctic therefore lacks Corydalidae.

In contrast, seven genera (six of Chauliodinae, one of Corydalinae) are known from the Nearctic, though none is boreal. Others, the neotropical *Platyneuromus* and *Chloronia*, abut the region in Mexico.

b) *Neotropical*. Both families are represented. The sialid fauna includes several species, which have been placed in *Protosialis* – itself a possible synonym of *Sialis*. One species from Chile may be more distinctive (p. 17).

Corydalidae include *Corydalus, Archichauliodes, Protochauliodes* and *Chloronia* (widely distributed), *Platyneuromus* (central America only) and *Nothochauliodes* (the latter from Chile only). The general ranges of *Chloronia* and *Platyneuromus* abut the southern parts of the Nearctic, as noted above.

c) *Ethiopian*. Only a few species of Megaloptera occur in the Ethiopian region, and appear to be restricted to southern Africa and Madagascar. Sialidae include the monotypic *Leptosialis* and two species of uncertain generic placement. Three chauliodine genera (*Platychauliodes, Madachauliodes* and *Taeniochauliodes*) and one corydaline genus (*Chloroniella*) are also present. The latter, isolated from all other Corydalinae, and the absence of Megaloptera from most of tropical Africa, are noteworthy features. ***Van der Weele*** (1906) noted a specimen of *Sialis* (as *Protosialis*) from Senegal but this has not been confirmed by any later record, and the record (of a New World species) was presumed erroneous (***van der Weele*** 1910).

d) *Oriental*. Sialidae are apparently rather sparse in the Region: the endemic *Indosialis* (Hong Kong) and a species of *Sialis* recorded from China (Szechwan) are the only taxa recorded. Corydalidae are much more diverse, with very characteristic genera of Corydalinae (*Acanthacorydalis, Neoneuromus, Neurhermes, Neuromus*) abutting the Palaearctic. The latter two, and *Protohermes*, extend southwards to the Indonesian archipelago. Only one chauliodine genus (*Anachauliodes*) is a regional endemic, whereas *Ctenochauliodes* and *Neochauliodes* enter the southeast Palaearctic. Recent studies by ***Yang & Yang*** and by ***Asahina*** are adding substantially to knowledge of the Oriental Corydalidae, and confirm that the Region is a major centre of diversity for this family.

e) *Australian*. Two small genera of Sialidae (*Austrosialis, Stenosialis*) are endemic to Australia. In the Chauliodinae, *Archichauliodes* is remarkably diverse, with 4 well-defined species groups in 2 subgenera, but only one species occurs in New Zealand where it is the sole megalopteron. Species of *Protochauliodes* (also in North America) occur in eastern Australia, and both these genera are also known from Chile.

Megaloptera are, apparently, absent from New Guinea and Sulawesi, Corydalinae from the whole Region, and Chauliodinae from Tasmania.

The distribution of Sialidae therefore includes a wide range of climatic regimes, from polar to tropical, whereas Corydalidae are more predominantly tropical, with outliers in various temperate areas. Corydalinae are the 'more tropical' subfamily and Chauliodinae rather more widespread. Two genera of Chauliodinae appear to reflect a southern (Gondwanan) origin, which possibly incorporates other Chauliodinae. The distribution of *Protochauliodes* may reflect movement of the group into the northern hemisphere. In South America, the Chilean fauna forms a distinct segregate: as in other groups of insects, the deserts to the north, the Andes to the east, and the extreme climate of the far south effectively isolate the fauna of its temperate-region taxa. *Sialis chilensis* appears to differ substantially from other South American Sialidae, and may merit generic distinction.

No genus of Megaloptera is world-wide, or close to it. *Sialis* is indeed widely distributed, though of uncertain integrity, but is absent from some very large parts of the world. Likewise, no genus is pan-tropical: indeed, no tropical megalopteran genus occurs in more than one major region. In the New World, the predominant genus *Corydalus* is very widely distributed.

## Ecology

Very few species of tropical Megaloptera have been studied in detail and, as noted earlier, many appear to be rather rare. Biological knowledge of the order has thus accrued predominantly from studies of temperate region forms. All have aquatic larvae, and there is no reason to suppose that any other major larval habitat will be found in the order. Most occur in permanent water although *Dysmicohermes* live in intermittent streams in high mountains of western North America (***Evans*** 1972), and this habitat has been noted also for several other Chauliodinae. Larvae of the New Zealand *Archichauliodes diversus* were once found in the ocean at the mouth of a stream (***Hamilton*** 1940), but they did not survive attempts to rear them in sea water, and the group appears to be entirely and obligately freshwater.

Within this, they range from hot springs at around 32 °C (*Corydalus cornutus* in Mexico) (***Lattin*** 1952, in ***Evans*** 1972) to cold mountain streams, in addition to swamps and ponds. The same species of Sialidae may occur in a wide range of habitats, ranging from small streams to lakes and large rivers (***Ross*** 1937), but many Corydalidae appear to be more specialised. Commonly, corydalid larvae are associated with more highly oxygenated 'riffle' areas, whereas sialid larvae occur in mud and bottom sediments.

Megaloptera larvae of certain temperate-region species can be very abundant. A maximum density of 815 *S. lutaria*/$m^{-2}$ was found in Essex (UK) (***Brooker*** 1979) and several other authors have reported densities of 70–>200/$m^{-2}$. Young *Corydalus* larvae can also occur at more than 200/$m^{-2}$ (***Brown & Fitzpatrick*** 1978). Values for production of *C. cornutus* in several sites in the Guadeloupe River, Texas (***Epperson & Short*** 1987) are amongst the highest reported for a single species of aquatic predatory insect, and were up to 22.9 g ash-free dry mass/$m^2$/year. Annual production varied considerably in streams with similar water quality but with different size, flow, and temperature (***Short et al.*** 1987), and this difference apparently reflected differences in growth and development resulting from differences in thermal regime. In Arkansas, where *C. cornutus* is univoltine, annual production/biomass ratios ranged from 7.4–9.0 in adjacent riffle habitats (***Bowles*** 1990). The abundance of *Sialis* larvae, in particular, suggests that they could exert considerable pressure on the benthic fauna (Fig. 18) (***Azam &***

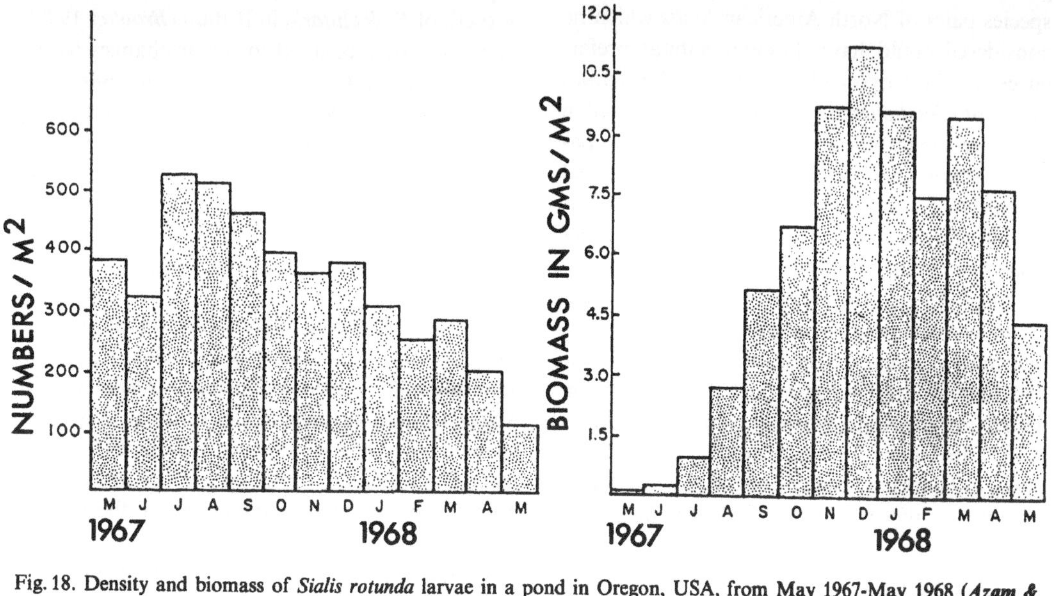


Fig. 18. Density and biomass of *Sialis rotunda* larvae in a pond in Oregon, USA, from May 1967-May 1968 (***Azam & Anderson*** 1969).

***Anderson*** 1969). Estimates of annual production of *Sialis* include 250 mg dry weight/$m^{-2}$ for *S. mitsuhashii* (extrapolated from ***Yamamoto*** 1972; ***Brooker*** 1979), 512 mg dry weight/$m^{-2}$ (***Giani & Laville*** 1973) or 985–1171 mg/$m^{-2}$ (***Brooker*** 1979) for *S. lutaria*.

Although generalisations on megalopteran biology may be unwise because of the strong bias towards studies of a limited number of temperate region taxa, it is inevitable that this risk be taken in an account of this sort. However, Sialidae studied in various parts of the world are overall rather similar in biology, and the strong framework available from work on North American Corydalidae has been augmented by work on east Palaearctic taxa.

## Habitats

a) Larvae. Although, as implied above, sialid larvae may at times occupy a wide spectrum of habitats, certain species appear to be more specialised. Thus, ***Evans*** (1972) found a western North American species, *S. arvalis*, only in warm streams.

Not all instars may have the same micro-distribution. Small *Sialis lutaria* larvae (first year of development) tend to be relatively more abundant in the littoral zones, and larger (second year) larvae more common in sub-littoral or profundal zones (***DuBois & Geigy*** 1935, ***Giani & Laville*** 1973).

Microhabitat differences probably occur between coexisting Sialidae, but opportunities to investigate this are limited. ***Ross*** (1937) listed several species pairs of North American *Sialis* which he considered could have different habitat preferences. In the Corydalidae, larvae of *Neohermes* and *Protochauliodes* are ecologically segregated in the same geographical area, sometimes by apparently small differences. ***Evans*** (1972) recorded *N. californicus* and *P. aridus* in the same Californian streams, where the former occurred in deeper faster water and the latter in shallower slower-moving regions. In Oregon, *N. californicus* coexists with *P. spenceri* in two small connected intermittent streams. These differed in substrate and only *N. californicus* occurred in the stream with a firm substrate, where *P. spenceri* was dominant in the softer substrate stream (***Evans*** 1972). ***Penny & Flint*** (1982) suggested the possibility that stream size may reflect specific requirements for species of *Chloronia*, as *C. hieroglyphica* never occurred near large streams or rivers in northern Brazil.

Western North American Chauliodinae exhibit a habitat dichotomy related to water temperature. ***Evans*** (1972) found several species only in cold water areas and others in warm water.

Two larvae of *Sialis joppa* have been recorded from inside pitchers of the eastern pitcher plant (*Sarracenia purpurea*) in New Jersey (***Mather*** 1981). Both were prepupal and, although no other larvae were found in nearby water, it is possible that they had entered the pitchers in search of pupation sites rather than having been resident there earlier in their lives.

The factors influencing larval habitat suitability are not wholly clear. Respiratory needs (p. 73) are clearly important for some Corydalidae (*Parachauliodes*: ***Hayashi*** 1989), as is water pH (*Sialis itasca*: ***Lilly et al.*** 1978; *S. aequalis*: ***Tarter & Woodrum*** 1972), and other aspects of water quality. *Nigronia fasciatus* can tolerate low pH values in the laboratory (***Butler et al.*** 1973), as can *Sialis aequalis* (***Tarter & Woodrum*** 1972), and more comprehensive details of chemical parameters of water bodies with *N. fasciatus* were summarised by ***Roback*** (1974). Whereas *Sialis* may at times tolerate acidic pollution (***Warner*** 1971), and larvae of two species of *Nigronia* in Georgia have been found in streams contaminated by textile industry wastes and sewerage treatment facilities, there are certainly interspecific differences in such tolerances (***Caldwell*** 1976). *N. serricornis* larvae are more tolerant to chemical extremes than are *N. fasciatus*. There was no evidence that herbicides (dalapon and 2,4-D i.e. 2,2-dichloroproprionic acid and 2,4-dichlorophenoxyacetic acid, respectively) used to control aquatic weeds had any effect on mortality or growth of *Sialis lutaria* in Britain (***Brooker*** 1979), although these resulted in major changes to the habitat. ***Takenaka*** (1975, cited by ***Hayashi*** 1989) claimed that *Parachauliodes japonicus* larvae can tolerate some water pollution, and their ability to use atmospheric oxygen in pollution-induced hypoxic conditions may be a factor in this (see p. 75). In shallow water they can readily raise their respiratory tubes to the water surface.

There is little specific reference to conservation of Megaloptera, but water quality is clearly a major determinant of their wellbeing. ***Asahina*** (1987) suggested that *Parachauliodes yanbaru*, apparently endemic to a small forested region of Okinawa Island, should be protected by conserving the forest and the streams inhabited by the larvae. The relative abundance of Megaloptera, together with most other insect orders except Diptera, decreased in stream communities subject to increased watershed development (Virginia, USA: ***Jones & Clark*** 1987).

b) Adults. All adult Megaloptera occur along watersides close to the corresponding larval habitat: they have only rarely been recorded elsewhere so that, in general, adult presence is a good guide to the larval habitat for the species concerned.

Putative mating swarms of adult males have occasionally been reported, for example in *Protochauliodes infuscatus* (***Smith*** 1970).

## Feeding habits

Larvae of Megaloptera have long been known to be voracious predators (***Davies*** 1903) and have been recorded eating a great variety of aquatic animals. *N. serricornis* and *Corydalus* larvae, and probably those of other species, forage with their mandibles held wide open and these would close on the prey. There was no evidence that the larvae could detect more distant prey (***Fuller & Hynes*** 1987). There is some possibility (***Smith*** 1970) that larval styli (67) may be used in sensing prey. Cannibalism is frequent, and probably occurs in all species at times: as examples, it is frequent amongst larvae of *Sialis rotunda* and *S. californica* (***Azam & Anderson*** 1969), occurs in *Neohermes californicus* and *Protochauliodes* (***Maddux*** 1952, ***Penland*** 1953), *Nigronia* (***Tarter et al.*** 1975) and in *Corydalus cornutus* (***Stewart et. al.*** 1973). The latter is also known to eat *Sialis* larvae and *Chauliodes* larvae. Cannibalism may be related to stress conditions (***Lilly et al.*** 1978). In captivity larvae can be induced to eat small pieces of beef (***Davies*** 1903). Chunks of beef about 1 cm diameter were devoured in about five minutes by large *Neohermes* larvae (***Smith*** 1970).

*Sialis* larvae generally feed on the most abundant members of the benthic invertebrate community, and are relatively indiscriminate. They consume small prey without chewing, but larger items are captured with mandibles and forelegs and the part fitting into the pharynx is severed (***Azam & Anderson*** 1969). In contrast to the generalist species studied by these workers, *S. itasca* preyed almost exclusively on ostracods even though many other possible prey were available (***Lilly et al.*** 1978) and *S. aequalis* had a relatively restricted range of available prey (***Woodrum & Tarter*** 1973). As specific feeding records, *S. lutaria* in Denmark fed on Oligochaeta, *Asellus* and Chironomidae (***Iversen et. al.*** 1985) and *S. rotunda* and *S. californica* in laboratory tanks ate larvae of chironomids, mayflies, caddisflies and damselflies, as well as tubificid worms (***Azam & Anderson*** 1969). Examination of gut contents or faeces has been used to determine sialid prey spectra.

Some corydalid larvae incorporate small amounts of algae or organic debris in their diets (***Cummins***, 1973) but it is probable that this represents only casual ingestion. A detailed study, from gut content analysis, of the diet of *Corydalus cornutus* by ***Stewart et al.*** (1973) revealed that major dietary items in Texas included *Simulium* larvae (25.8 % frequency of 468 stomachs examined) and the trichopteran larvae *Cheumatopsyche* (23.9 %) and *Hydropsyche* (19.0 %), and a range of 19 other food groups. Collectively, Ephemeroptera constituted 7.2 % of the diet, Trichoptera 49.7 % and Diptera 32.7 %. Different size classes of larvae ingested generally similar diets, and *C. cornutus* accepted most organisms present in its environment, with possible exceptions of some taxa of Trichoptera and Odonata. Analysis of gut contents of *Nigronia serricornis* larvae in Ontario (***Fuller & Hynes*** 1987) showed that net-spinning Trichoptera (mainly Philopotamidae and Polycentropodidae) were predominant prey items except in summer, when the diets were more diverse (Fig. 19). In January and February, all 35 larvae collected had empty guts. Serological tests to determine the predators of *Simulium fulvinotum* in Brasil, showed that 11 of 16 corydalid larvae tested had fed on this prey (***Gorayeb & Pinger*** 1978).

*Protohermes grandis* in Japan fed on Ephemeroptera, Trichoptera and Chironomidae (***Yoshida et al.*** 1985). All of these were eaten also by *Nigronia fasciatus* in West Virginia (***Tarter et al.*** 1975) but the isopod *Asellus aquaticus* was the most frequent food item, ranging from 22.9 % (summer) to 56.0 % (spring). Food of *Neohermes californicus* included mayfly and caddis larvae, sialids, and tadpoles (***Smith*** 1970). The type of substrate (very coarse sand, gravel/pebble, artificial turf) influences the susceptibility of prey to *Nigronia* by the availability of prey refuges and the ease with which the predator can sense or pursue prey (***Fuller & Rand*** 1990). Laboratory studies of prey suitability were undertaken also by ***Fuller & DeStaffan*** (1988).

For *P. grandis*, the maximum size of prey eaten increased with larval size, and large larvae did not take the smallest prey (***Hayashi*** 1988), even though these were abundant. The larvae are nocturnally active and are usually ambush predators (***Hayashi & Nakane*** 1988). They are tactile predators and capture prey with their mandibles

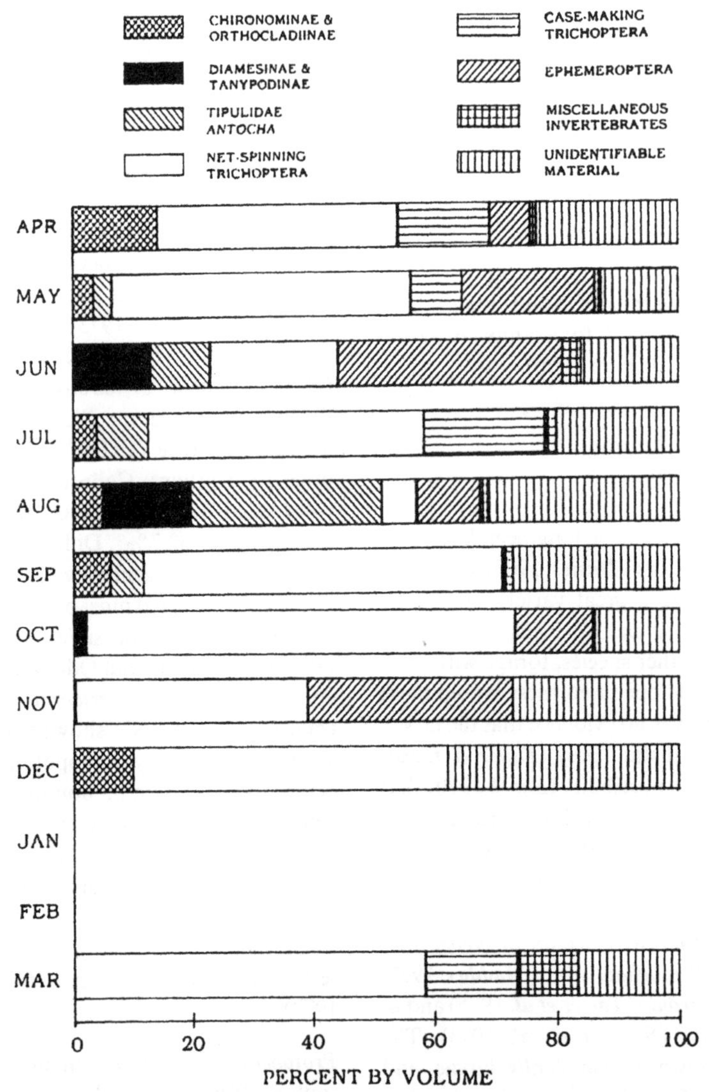


Fig. 19. Seasonal variation in prey spectrum of *Nigronia serricornis* larvae in Ontario: percentage by volume of gut contents in monthly samples; key to main prey types given at top of figure (April 1981 – March 1982) (***Fuller & Hynes*** 1987).

(***Hayashi*** 1985). ***Hayashi*** undertook a detailed study of the handling time of this species, and found this to be affected by water temperature, shape and hardness of prey, but not by hunger. In a pioneering study, ***Hayashi & Nakane*** (1988) attached small radio transmitters to *Protohermes* larvae and tracked their foraging movements. Attachment of transmitters appeared not to affect larval behaviour adversely, and some larvae moved actively by night (Fig. 20), with the degree of movement appearing not to be hunger-dependent.

Adult Megaloptera are rather short-lived, and there is strong suggestion that they do not feed but depend on larva-accumulated reserves for their reproductive maturation. *S. lutaria* adults have been found with adherent pollen, so that there has been supposition that adults visit flowers to feed (***DuBois & Geigy*** 1935). They appear not to chew food, but may at times squeeze flower organs between their mandibles and extract nectar. ***Parfin*** (1952) noted that *C. cornutus* and *Chauliodes*

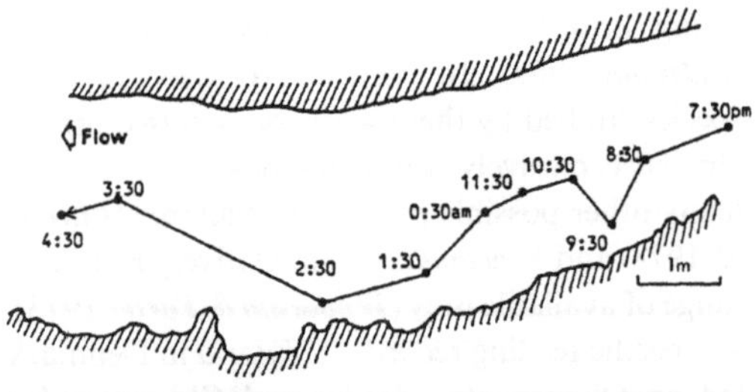


Fig. 20. The movement of a larva of *Protohermes grandis* fitted with a radiotransmitter. Times (from 7.30 pm) are shown (***Hayashi & Nakane*** 1988).

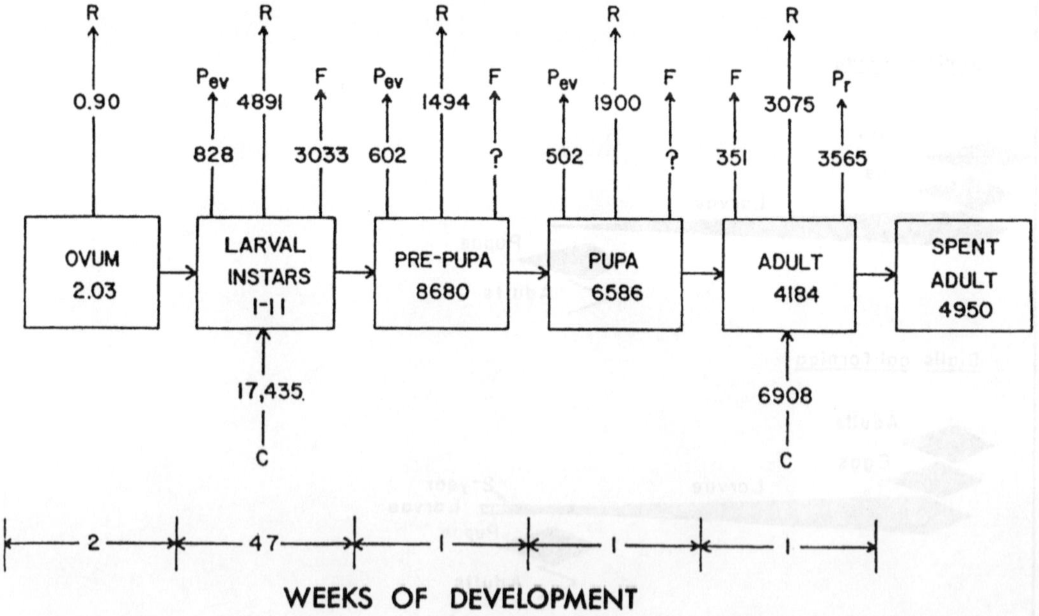


Fig. 21. Energetics of development of a typical female individual of *Corydalus cornutus*. Figures are in Joules for C, consumption; R, respiration; F, egestion; Pev, production of exuviae; Pr, production of egg masses. For the larva (47 weeks) assimilation (A) is 14,401 joules, growth (Pg) 8,682. Male pupal respiration was 2130 joules/week, and male adult respiration, 2615 joules/week. Average adult female minimal reproductive effort was 2791 joules (eggs) and 774 joules (egg cases). (***Brown & Fitzpatrick*** 1978).

adults in captivity 'eagerly consumed' a honey-water solution, although maple syrup-water solutions were rejected. Fermented molasses with water and sugar and similar baits (including fermented bananas) were earlier known to be attractive to adults of these genera. The adults of *Protohermes grandis* have been seen to feed on tree sap (***Yoshida et al.*** 1985). For most taxa, no information on adult feeding habits is available. ***Balduf***'s (1939) summary of Sialidae biology included the comment that, although nectar may not be essential to adult life, it 'certainly prolongs life materially'.

Consumption rates of *C. cornutus* larvae were measured, using *Glyptotendipes* chironomids as prey, and metabolism of female adults also investigated as part of a study of population energetics of this species (***Brown & Fitzpatrick*** 1978). Assimilation efficiency (83 %) is of a similar value to that of some other invertebrate predators studied and overall production efficiency was high. About half (50.1 %) of inferred adult female consumption was allocated to reproduction, and the adult energy intake was about one third of the total larval energy intake (Fig. 21). Whereas females ingested sugar water or fruit juices in the laboratory, males drank only water: a sexual difference which may account for the variability in some earlier observations cited above, in which this factor was not, apparently, considered.

Males of *C. cornutus* were apparently impeded from feeding by their large mandibles. Calorific values of all major life history stages, together with measurements of consumption (inferred for the adult), egestion, assimilation, respiration, growth, and production of exuviae and egg masses were used by ***Brown & Fitzpatrick*** to assess the energy budget of 'typical' individual *C. cornutus* and the population energetics of this species. The clear implication from their data (Fig. 21) is that females must consume food, as the energy content of gravid females did not differ significantly from recently emerged or 'spent' individuals.

## Lifehistories and voltinism

Although adult Megaloptera tend to be short-lived, the converse is true for larvae, so that the overall lifehistories of the insects tend to be long. The regular time of appearance of *Sialis* adults in temperate regions indicates that they have a univoltine lifecycle but, as ***Davis*** (1903) and ***Dubois & Geigy*** (1935), amongst others, have implied, larvae persist for at least two years. Thus the European *S. lutaria* has a two-year lifecycle in

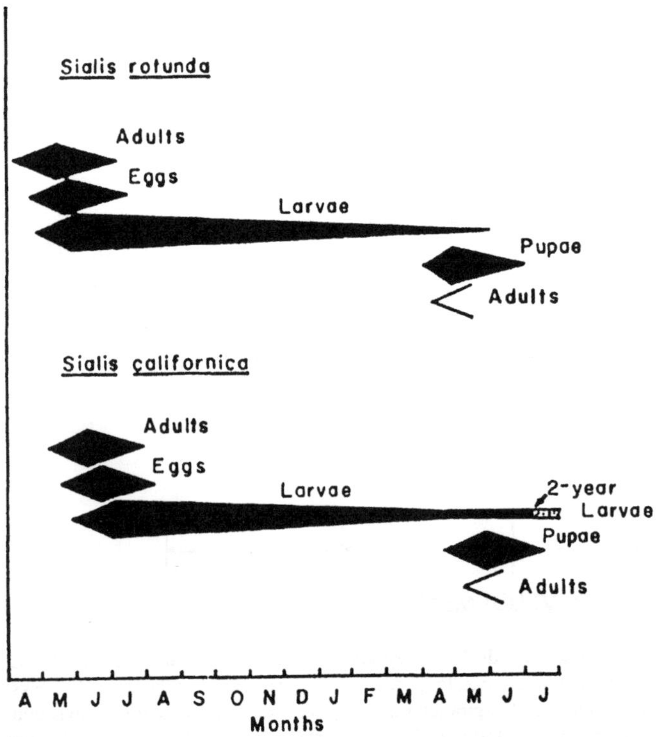


Fig. 22. Life history of *Sialis*: *S. rotunda* and *S. californica* in Oregon, USA (***Azam & Anderson*** 1969).

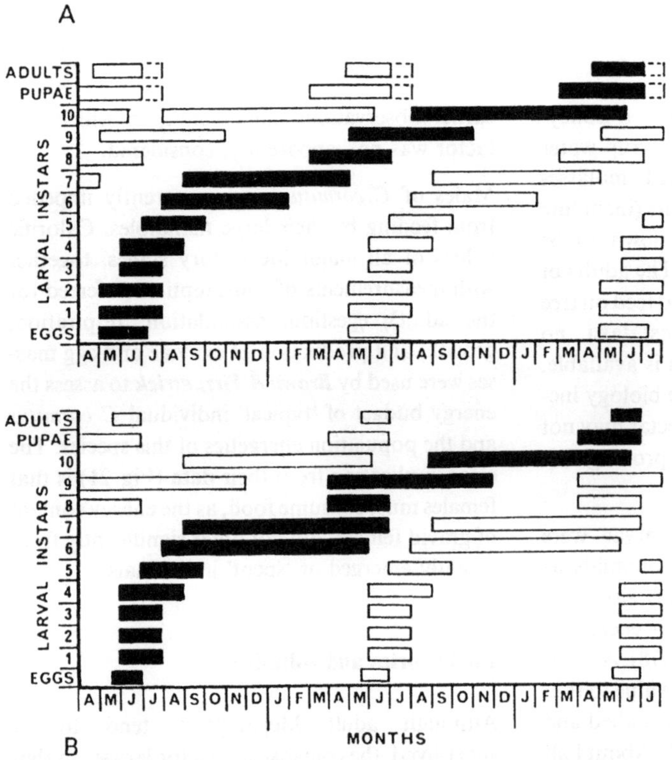


Fig. 23. Life history of *Sialis*: *S. lutaria* (A) and *S. fuliginosa* (B) in Britain: one sequence of cohorts for each in black, months indicated by initial letters (***Elliott*** 1977).

Britain (Fig. 23) (***Elliott*** 1977, ***Brooker*** 1979), but takes up to three years to complete development in a high altitude lake in the Pyrenees (Fig. 24) (***Giani & Laville*** 1973). Whereas this trend at high altitudes may be widespread, lower altitudes (with generally warmer water temperatures) result in a two-year cycle (***Geigy & Grobe*** 1958). In contrast, it has a one-year cycle in parts of Denmark (***Iversen & Thorup*** 1987). However, *S. rotunda* in North America is indeed univoltine (Fig. 22)

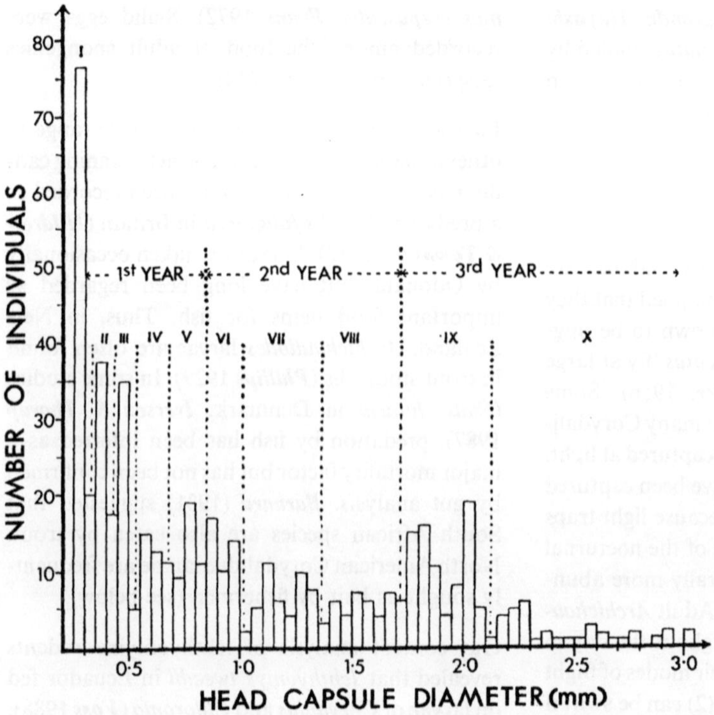


Fig. 24. Seasonal development of *Sialis lutaria.* Cohorts (instars) indicated by larval head diameter (mm) over three years in a population studied at 2285 m (central Pyrenees) (***Giani & Laville*** 1973). Compare with the shorter life cycle for the same species shown in Fig. 23.

(***Azam & Andersen*** 1969), whereas *S. californica* (a larger species) can take either one or two years to develop, depending on time of oviposition and availability of prey. The larval stage of *S. rotunda* lasts 10–11 months, and most larvae of *S. californica* also take only one year to develop. Eggs of both species hatch within about 8–12 days (***Azam & Anderson*** 1969).

Early observations on North American Corydalidae (***Davies*** 1903) implied that their larval development took three years, and later work suggests that this may be even longer under unfavourable conditions. ***Evans*** (1972) believed that at least four years were needed for development of *Dysmicohermes disjunctus*, and larvae of *D. ingens* maintained in the laboratory moulted only twice in three years. Five years may be needed by *Orohermes crepusculus* in cold mountain streams, and the same period for many larvae of *Neohermes californicus*. In Japan, *Protohermes grandis* takes up to three years to develop (***Yoshida et al.*** 1985), but larval duration reflects water temperature (***Hayashi*** 1988 b). In contrast, several univoltine populations of *C. cornutus* have been reported (***Bowles*** 1990, ***Brown & Fitzpatrick*** 1978).

Although larval development normally proceeds without delay after hatching from the eggs, some taxa frequenting more temporary habitats (such as seasonal streams) may aestivate after eclosion. This appears to be usual with *Protochauliodes aridus* and, probably, with *Neohermes californicus*, for examples (***Smith*** 1970). Larvae of *P. aridus* burrowed into the gravel of dry stream beds (the females may deposit their eggs on stones in the dry beds) to a depth of 10–30 cm where they made small cells beneath stones and remained until the streams were again full. Aestivation is not obligatory, but some moisture is necessary for the larvae to survive: in a completely dry environment they died in three or four days (***Penland*** 1953). ***Smith*** (1970) suspected that this kind of lifecycle could be more widespread in Californian Corydalidae.

The last instar (10) of *Sialis* is commonly the overwintering stage, which does not develop further until spring. Despite complications afforded by long life cycles in at least some taxa, seasonal patterns tend to be reasonably well-defined in temperate region Megaloptera. Thus, for a univoltine population of *C. cornutus*, ***Brown & Fitzpatrick*** (1978) could state confidently that the 'typical' larva hatched in mid-June, grew rapidly until November and again in March, reaching adult size prior to pupation in early June, with a 'typical' larval life of 331 days. Likewise, *Sialis* phenological patterns can be plotted relatively precisely, with characteristic differences between species, in a given locality and, even though the overall duration of development may differ substantially in relation to latitude or altitude, characteristic cohorts are present at a given time (Fig. 24). Local variations in water temperature can influence duration of development (***Short et al.*** 1987), as in *P. grandis* (***Hayashi*** 1988b). One population of *C. cornutus* studied by them in Texas may even be bivoltine, a condition not reported for any other megalopteron.

## Dispersal

As adult Megaloptera, particularly Sialidae, are rarely found far from water, it is implied that they disperse rather little. None is known to be migratory but *Protochauliodes infuscatus* 'fly at large over the countryside' (***Chandler***, 1956). Some Sialidae are generally diurnal, but many Corydalidae are nocturnal and have been captured at light. Likewise, Australian Sialidae have been captured in numbers at lights. Possibly because light traps may give a collecting bias, some of the nocturnal Megaloptera appear to be generally more abundant than many diurnal species. Adult *Archichauliodes* can be on the wing throughout the day in dull weather (***Riek*** 1970), and their modes of flight vary. It (1) is normally slow but (2) can be swift if the insects are disturbed and (3) can constitute gliding high above the streams. In the latter the wings are outspread, raised above the horizontal and slightly swept back. Both sexes do this, and the function is unknown.

Megaloptera have occasionally been categorised as 'weak' or 'feeble fliers', but Corydalidae are considered to be stronger than Sialidae (***Davis*** 1903).

Likewise, most larval Megaloptera appear to move around rather little, although this has only rarely been quantified or commented upon. ***Davis*** (1903) commented on the low activity of young *Chauliodes* larvae. Radio-tracking of *Protohermes grandis* larvae (***Hayashi & Nakane*** 1989) revealed two distinct phases of activity, a stationary phase and an active one. Many larvae remained in the same place for several days, but they may become more active at night. The presence of megalopteran predators may influence the choice of substrate by prey, to increase chances of avoiding mortality from predation: for example in *Hydropsyche* (Trichoptera) (***Michael & Culver*** 1987).

## Natural enemies

Many different kinds of predators attack Megaloptera, although specific records are sparse, and all stages can be vulnerable. Corydalid eggs are eaten by anthicid beetle larvae (*Anthicus haldemani* on *C. cornutus* in North America: ***Howard*** 1896), and slugs (*Arion circumscriptus* on *Orohermes crepusculus* ***Evans*** 1972). Sialid eggs were recorded among the food of adult spongeflies (Sisyridae) by ***Tjeder*** (1944).

Larvae are likely to be taken by a wide range of other aquatic invertebrates. The net-spinning caddis *Plectrocnemia conspersa* has been recorded as a predator of *Sialis fuliginosa* in Britain (***Hildrew & Townsend*** 1982). Larvae are taken occasionally by Odonata, but have long been regarded as important food items for fish. Thus, in New Zealand, *Archichauliodes* larvae are often found in trout stomachs (***Phillips*** 1929). In some studies (*Sialis lutaria* in Denmark, ***Iversen & Thorup*** 1987), predation by fish has been inferred as a major mortality factor but has not been confirmed by gut analysis. ***Barnard*** (1931) suggested that South African species are also eaten by trout. North American Corydalidae larvae are frequently sought as bait by fishermen (see below).

Gut content analysis of ichthyomyine rodents revealed that *Ichthyomys tweedii* in Ecuador fed on larvae of *Corydalus* and *Chloronia* (***Voss*** 1988). These larvae comprised a substantial proportion of the diet and even the largest larvae (up to 65 mm long) were eaten, and are chewed into fragments by the incisors. The related *I. pittieri* also kills dobsonfly larvae and tears them into fragments with incisor bites.

Loose wings of *Nigronia* scattered along stream banks imply that birds or other predators attack the adults (***Tarter et al.*** 1975). Adult head capsules of *Dysmicohermes* have been found in pellets of the spotted owl, *Strix occidentalis* (***Evans*** 1972). Bird predation on adults may be more common than implied from these sporadic records.

Several possible defensive mechanisms against predators have been suggested. *Neohermes* larvae and prepupae, for example, vomit a noxious-smelling liquid when disturbed (***Smith*** 1970), and larvae also exhibit a very pronounced avoidance reaction on encountering others. Adult female *Corydalus* can bite (***Parfin*** 1952).

Parasites of eggs include *Trichogramma evanescens* (***Silvestri*** 1908), and *T. semblidis* (***Azam & Anderson*** 1969). The latter attacked some 14 % of egg masses of *Sialis rotunda* (mainly peripheral eggs) and 37–65 % of masses of *S. californica* (often most of the eggs). The identities of these two parasites have historically been confused, but they were considered discrete by ***Salt*** (1937). Differences in parasitisation rates of different host species may reflect different oviposition patterns – eggs of *S. californica* have more of the egg surface exposed for parasite oviposition, but those of

*S. rotunda* have only the end surface exposed to attack.

The biology of *T. semblidis* parasitising *S. lutaria* was studied in detail by ***Salt*** (1937). Of nearly a quarter of a million host eggs (totalling 386 egg-masses), about 0.6 percent (in 19 batches) were parasitised. The male parasite is dimorphic, and those reared from *Sialis* were apterous; individuals reared from eggs of three species of Lepidoptera were winged. Eggs containing parasites turn black, and eggs which have yielded adult parasites are easily detectable by having large round lateral holes.

Small larvae tend to be much more abundant than large larvae in estimates of standing crop in field populations (*Sialis*: ***Azam & Anderson*** 1969, ***Pritchard & Leischner*** 1973; *Corydalus*: ***Brown & Fitzpatrick*** 1978), implying that overall larval mortality may be very high. Fungus attack has been noted on eggs and larvae and there are sporadic occurrences of disease-like symptoms which have not been clarified further. ***Smith*** (1970) found that male pupae (but not females) of *Neohermes californicus* died from an apparent disease which resulted in their wingpads becoming filled and distended with haemolymph, for example.

## Other associations

Phoretic associations of larval Chironomidae and of Entoprocta have been recorded with corydalid larvae in North America (***Hilsenhoff***, 1968, ***Benedict & Fisher*** 1972, ***Gotceitas & Mackay*** 1980, ***Furnish et al.*** 1981, ***Tracy & Hazelwood*** 1983). The entire larval life of *Plecopteracoluthus downesi* (now *Nanocladius*) is passed on *Chauliodes pectinicornis* larvae, where the fly larvae feed on debris accumulating on the host surface. More than 90 percent of hosts collected in New Hampshire supported symbionts, with about one third of the larvae inhabiting the mesothoracic sternal area and others, the abdominal tergites (***Benedict & Fisher*** 1972). The midge occurred also on other hosts, including *C. cornutus* and, although most larvae were found close to the gills by ***Tracy & Hazelwood*** (1983), attachment sites were varied. Late instar *Nanocladius rectinervis*, in contrast, appeared to be specific on *Nigronia serricornis* in Ontario (Fig. 25) (***Gotceitas & Mackay*** 1980). The entoproct *Urnatella gracilis* was found near a gill base of *C. cornutus* in Missouri (***Tracy & Hazelwood*** 1983).

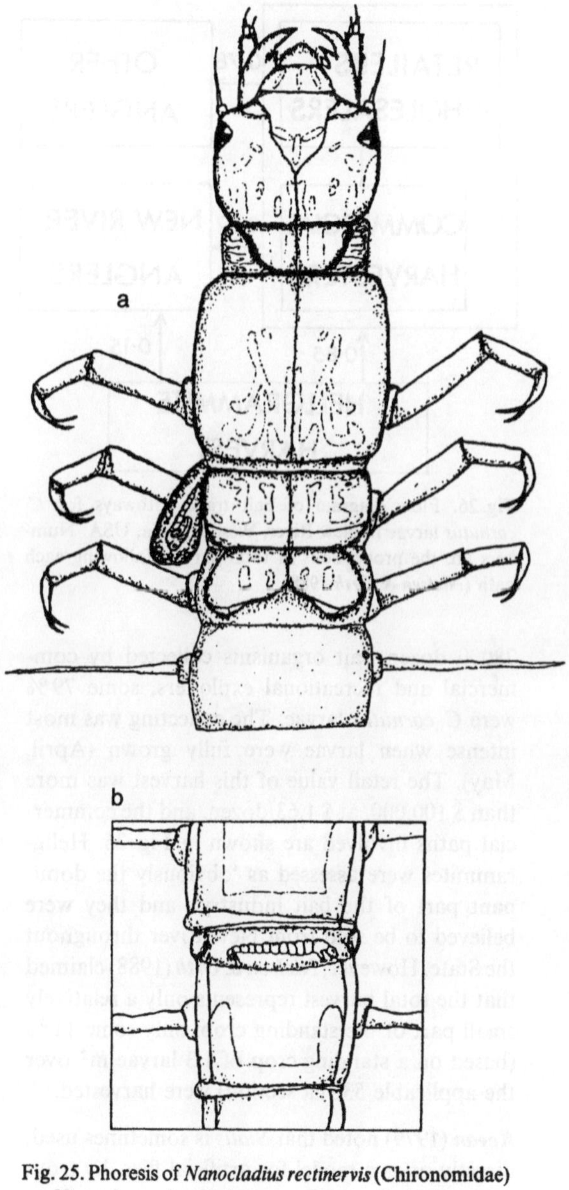


**Fig. 25. Phoresis of *Nanocladius rectinervis* (Chironomidae) on *Nigronia serricornis* larvae: a, pupa attached to host thorax; b, larva on seventh abdominal sternite of host (*Gotceitas & Mackay* 1980).**

## Economic importance

For most taxa, no economic importance is evident. There are two ways in which Megaloptera are sometimes stated to have positive value: as the predominant foods of commercially or recreationally-desirable fish, and as the basis of local supplies of fishing bait. Much of the latter relies on larvae of *Corydalus cornutus* ('hellgrammites') in North America, and a recent appraisal of the bait industry in West Virginia (***Nielsen & Orth*** 1988) indicates the extent of exploitation. In the New River and its tributaries, anglers collected 9153 dozen hellgrammites in 1983 and, of the total

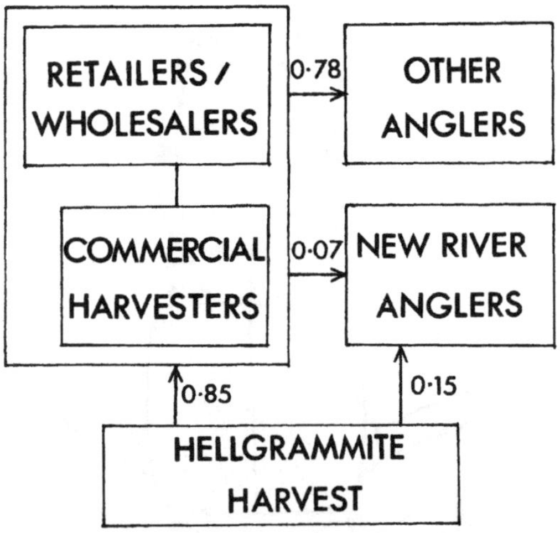


Fig. 26. Flow diagram of bait trade pathways for *C. cornutus* larvae in New River, West Virginia, USA. Numbers are the proportions of total harvest following each path (***Nielsen & Orth*** 1988).

78000 dozen bait organisms collected by commercial and recreational exploiters, some 79 % were *C. cornutus* larvae. The collecting was most intense when larvae were fully grown (April, May). The retail value of this harvest was more than $ 100,000, at $ 1.63/dozen, and the commercial paths involved are shown in Fig. 26. Hellgrammites were assessed as 'obviously the dominant part of the bait industry', and they were believed to be sold from New River throughout the State. However, ***Nielsen & Orth*** (1988) claimed that the total harvest represents only a relatively small part of the standing crop: only some 11 % (based on a standing crop of 1.3 larvae/m$^2$ over the applicable 536 ha section) were harvested.

***Kevan*** (1979) noted that *Sialis* is sometimes used, directly or as a model for artificial flies, by trout fisherman.

Dried larvae of *Protohermes grandis* in Japan ('magotaro mushi') are considered a remedy for emotional irritation of infants (***Kuwayama*** 1962).

*Sialis* larvae in Europe have been implicated in transmission of a hookworm parasite of fish, *Neoechinorhynchus rutili* (Acanthocephala). An early report (***Villot*** 1885) indicated larval hookworms apparently encapsulated in the fat body of *Sialis*, but they were found free and unencapsulated in the haemocoel of *S. lutaria* larvae by ***Lassiere*** (1988). Brown trout (*Salmo trutta*) infested with *N. rutili* were found to be feeding on *Sialis* larvae carrying worms and experimental infection of commercial rainbow trout (*Salmo gairdneri*) by feeding with such larvae was successful and may reflect a natural route of parasite transmission. *S. lutaria* larvae (whose infection rate was 7 % of 215 larvae captured, all those infected being of head capsule diameters in the range 1.92–2.67 mm) were believed to be an additional intermediate host, ingesting the worms in ostracod hosts.

### Cleaning behaviour

***Jander***'s (1966) comparative study of grooming behaviour included observations on *Sialis*, and some possible phylogenetic implications were revealed. As in Trichoptera, but unlike Planipennia, the antennae were cleaned together between raised forelegs (Fig. 27). Wing-grooming was also lacking in Planipennia, but *Sialis* cleans the wings with the abdomen, which is moved laterally between the forewing and hindwing. This probably occurs also in *Raphidia*.

## Adult Morphology

Knowledge of general morphology of the Megaloptera is largely extrapolated from detailed accounts of very few species. *Corydalus* and *Sialis* have both been examined in considerable detail, and accounts of their structure, together with that

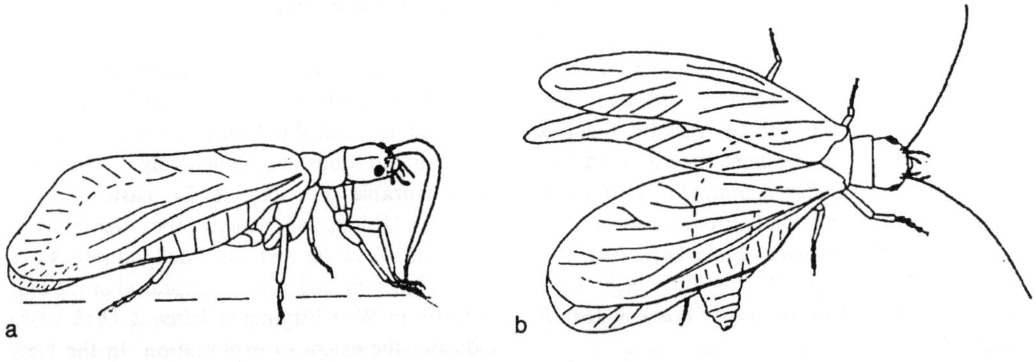


Fig. 27. Cleaning behaviour in *Sialis fuliginosa*: a, antennal cleaning by the forelegs; b, abdomen moved laterally between wings (***Jander*** 1966).

of *Chauliodes*, provide the basis for much generalisation and for the summary accounts published in many general entomology texts. They are here augmented, to illustrate the diversity of structure in the order, from the more detailed data presented in many taxonomic papers. Illustrations in the Systematic Account augment those in the following pages. Anatomical details also rely heavily on accounts of *Corydalus* and *Sialis*, representing major groups in the order.

Few teratological specimens have been recorded for any stage, although a bifurcated hind tibia of a young larval *C. cornutus* was reported by ***Wickham*** (1903).

## Head

The head of Megaloptera is relatively unmodified. The head capsule of several species has been described in considerable detail, but three accounts are particularly notable: ***Röber*** (1941) on *Sialis flavilatera* (now *S. lutaria*), ***Maki*** (1936) on *Chauliodes formosanus*, and ***Kelsey*** (1954) on *Corydalus cornutus*, these together exemplifying the three major segregates in the order. A broader comparative study of the corydaline head was provided by ***Glorioso*** (1981).

In all, the head (Fig. 28) is heavily sclerotised; in Corydalinae it is more conspicuously progna-

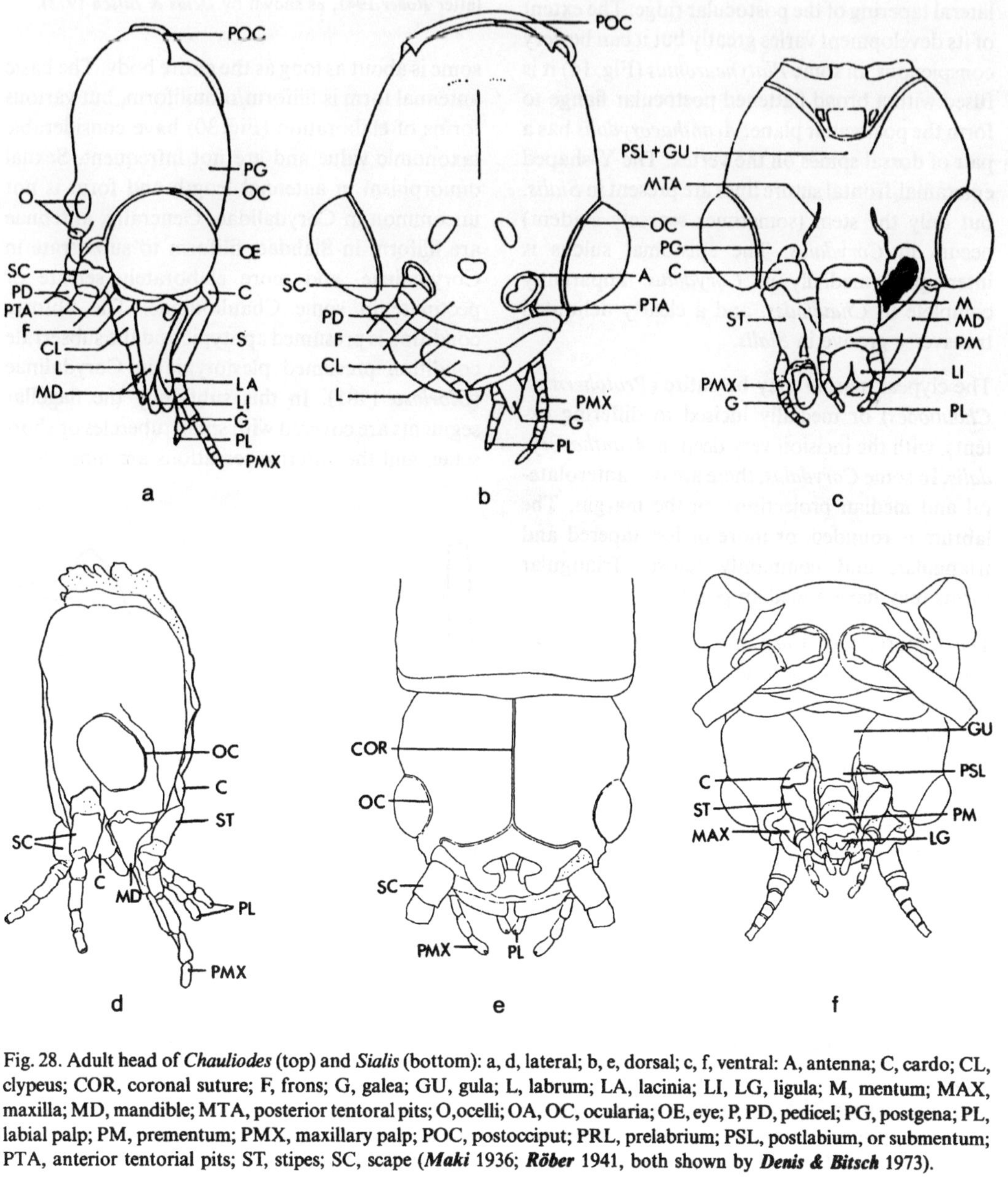


Fig. 28. Adult head of *Chauliodes* (top) and *Sialis* (bottom): a, d, lateral; b, e, dorsal; c, f, ventral: A, antenna; C, cardo; CL, clypeus; COR, coronal suture; F, frons; G, galea; GU, gula; L, labrum; LA, lacinia; LI, LG, ligula; M, mentum; MAX, maxilla; MD, mandible; MTA, posterior tentoral pits; O, ocelli; OA, OC, ocularia; OE, eye; P, PD, pedicel; PG, postgena; PL, labial palp; PM, prementum; PMX, maxillary palp; POC, postocciput; PRL, prelabrium; PSL, postlabium, or submentum; PTA, anterior tentorial pits; ST, stipes; SC, scape (***Maki*** 1936; ***Röber*** 1941, both shown by ***Denis & Bitsch*** 1973).

thous and flattened than in other major groups. In most genera of Corydalinae the head is noticeably flattened, but is more robust, and less flattened, in *Protohermes* and *Neurhermes*. The compound eyes are large and the ocelli also large in Corydalidae. Ocelli are close together in most genera of Corydalinae. The median ocellus generally has its posterior margin at about the level of the anterior margin of the lateral ocelli, but is distinctly anterior to them in most *Neurhermes* and some *Protohermes* (***Glorioso*** 1981). The ocelli are universally lacking in Sialidae, which commonly have small depressed areas in equivalent positions.

The postocular ridge of Corydalinae extends laterally behind the eyes to the occiput (***Glorioso*** 1981), and the postocular spine arises from the posterior angles of the head, where it forms a lateral tapering of the postocular ridge. The extent of its development varies greatly but it can be very conspicuous: in some *Platyneuromus* (Fig. 14) it is fused with a broad flattened postocular flange to form the postocular plane. *Acanthacorydalis* has a pair of dorsal spines on the vertex. The Y-shaped epicranial/frontal suture lines are present in *Sialis*, but only the stem (sometimes scarcely evident) occurs in *Corydalus*. The epistomal sulcus is interrupted medially in *Corydalus*, apparently complete in *Chauliodes*, and a clearly delimited transverse groove in *Sialis*.

The clypeal margin may be entire (*Protohermes, Chauliodes*) or medially incised to differing extents, with the incision very deep in *Acanthacorydalis*. In some *Corydalus*, there are two anterolateral and median projections on the margin. The labrum is rounded, or more or less tapered and triangular, and commonly setose. Triangular forms may have a median point.

The tentorium of *Corydalus* has no bridge between the arms of opposite sides of the head, and the functions of the bridge are assumed by the ventral arms (***Kelsey*** 1954). However, the tentorium is complete in *Sialis*, where the pretentorial arms subtend the antennae at the ends of the epistomal suture. The supratentorial arms extend between the antennae and the eyes (Fig. 29). The body of the *Sialis* tentorium has a long posterior projection, which is already present in the larva. The metatentorial arms are displaced anteriorly, reflecting prognathy.

*Antennae*. The antennae are long, commonly about as long as the head and prothorax together, with about 30–85 flagellomeres. The number of segments, within a broad range, is quite characteristic for particular genera. In male *Corydalus* the antenna can be considerably elongated and in

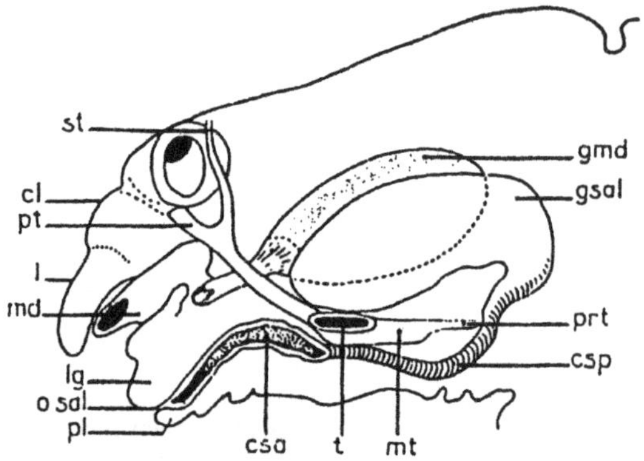


Fig. 29. *Sialis* head and tentorium: cl, clypeus; csa, anterior part of salivary canal; csp, lateral (paired) part of this canal; gmd, mandibular gland; gsal, salivary gland; l, labrum; lg, ligula; md, mandible; mt, metatentorium; osal, salivary orifice; pl, prelabium; prt, posterior process of tentorium; pt, pretentorium; st, supratentorium; t, tentorial body (after ***Röber*** 1941, as shown by ***Denis & Bitsch*** 1973).

some is about as long as the entire body. The basic antennal form is filiform/moniliform, but various forms of elaboration (Fig. 30) have considerable taxonomic value and are not infrequent. Sexual dimorphism in antennal length and form is not uncommon in Corydalidae. Generally, antennae are filiform in Sialidae, filiform to subserrate in Corydalinae, and more elaborately serrate or pectinate in some Chauliodinae. The filiform condition is presumed apotypic and the subserrate condition presumed plesiotypic in Corydalinae (***Glorioso*** 1981). In this subfamily the flagellar segments are covered with small tubercles or short setae, and the anterior serrations are most deve-

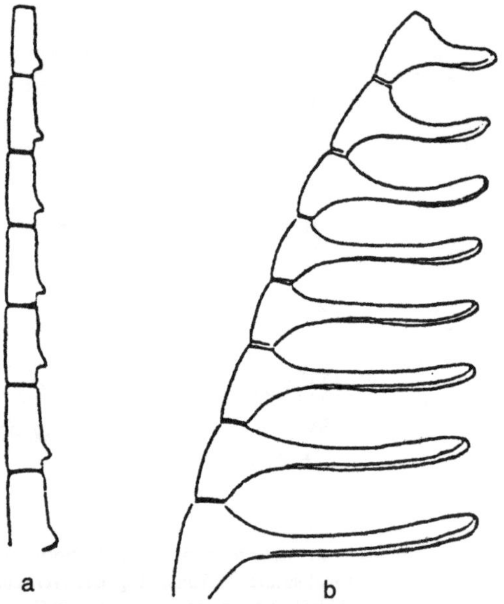


Fig. 30. Antennae: flagellar segments of *Corydalus* (a) and *Neochauliodes sinensis* (b) males (***Stitz*** 1926).

loped in *Neurhermes* and *Protohermes*. Pectinate antennae are rare in females, but occur in *Ctenochauliodes*.

The scape and pedicel are cylindrical and, essentially, unmodified. The scape articulates in the antennal socket by means of a median dorsal process.

*Mouthparts*. Mouthparts are 'basic mandibulate' in form. The mandibles are typically small and triangular with two medial teeth in Sialidae and Chauliodinae, but are usually more elongate and with three teeth along the medial margin in Corydaline. They are usually shorter than the head but are sometimes greatly enlarged in males (*Acanthacorydalis, Corydalus*: to about 30 mm long) and in both these genera the mandibles may be long 'tusks' and as long as the entire body. The large mandibles have led to suggestion that the insects may be extremely dangerous (***McLachlan*** 1899). Medial mandibular teeth are lacking in *Corydalus* and reduced (usually with the most basal tooth evident) in *Acanthacorydalis*. Female *Acanthacorydalis* may also have mandibles moderately enlarged, to be about as long as the head. The mandibles articulate near the anterior margin of the eye and may be rugose over much of their surface.

In the maxilla, the cylindrical maxillary palpi (usually 5-segmented, in Corydalidae but 4-segmented in *Corydalus*; 4-segmented in Sialidae) arise from a distinct palpifer. The apical segment is rounded or conical and has one or two membranous areas with dense microsetae (***Glorioso*** 1981), and the preapical segment has a distal annulus of alternating longer and shorter setae. Palpi of Sialidae have a subapical conical sensillum. In Corydalinae, the maxilla articulates at the side of the submentum.

The cardo is triangular and the stipes more or less cylindrical and 1 ½–2 times as long as broad. The lacinia is triangular and fused with the stipes. It is usually setose and, in many genera, has three especially long apical setae. ***Glorioso*** (1981) noted that the galea and stipes are separated by membrane and two horizontal subgaleal sclerites, the basal one often divided by membrane. The galea itself is densely setose (with long setae flattened and lanceolate in *Neurhermes* and *Protohermes*) and the apex usually has a small sensory 'peg' (absent in *Neurhermes*, weakly developed in *Neuromus, Neoneuromus*: ***Glorioso***) with apical microsetae. In Sialidae the galea is short, but it is longer and tapered in Corydalidae.

The labial palpi are 3- or 4-segmented, and usually similar in form to the maxillary palpi, although the apical segment is triangular in *Corydalus*. The ligula is membranous and bilobed. The mentum is usually not heavily sclerotised basally, and the prementum is short. The gular plate is long and conspicuous, laterally fused to the cranium and sometimes emarginate at the apex. It may be T-shaped (expanded at the apex) and is referred to as the gulamentum.

## Cervical region

The cervical region is considerably longer in Corydalidae than in Sialidae and, apparently, the origin of the lateral cervical sclerite differs in the two families. In *Corydalus* (***Crampton*** 1926, ***Kelsey*** 1954) and *Chauliodes* (***Maki*** 1936), this sclerite comprises a large sclerotised ventral plate and a pair of less sclerotised dorsal plates (together comprising the 'jugulare' of ***Crampton***, 1926), all anterior to the preepisternum + basisternum of the prothorax, and is derived from the presternum. In *Sialis*, the lateral cervical sclerite is the preepisternum which has become detached narrowly from the katepisternum and is connected anteriorly with the ventrolateral angles of the postocciput.

Thus, *Sialis* resembles Planipennia in this feature, whereas the condition in Corydalidae resembles that in many hemimetabolous insects in having the lateral cervical sclerite formed from the presternum, whilst the condition of the preepisternum being detached from the episternum occurs also in many other Holometabola.

*Chauliodes* and *Corydalus* have a small median unpaired plate with a smaller sclerite each side of this behind the ventral posterior margin of the head. The dorsal cervical sclerite is present in *Corydalus* (***Kelsey*** 1954).

## Thorax

The pronotum (Fig. 31) is elongate and posteriorly widened in Corydalidae but usually more quadrate and transverse (up to about twice as wide as long) in Sialidae. The main dorsal area is the protergal plate which is arched in Corydalidae to cover part of the pleural area. The anterior of the plate is less thickened and nearly continuous with the dorsal cervical membrane. Posteriorly there is also a transverse less thickened region. The ventral side of the prothorax includes a large undivided sclerite, often termed a basisternum, which extends posteriorly between the forelegs.

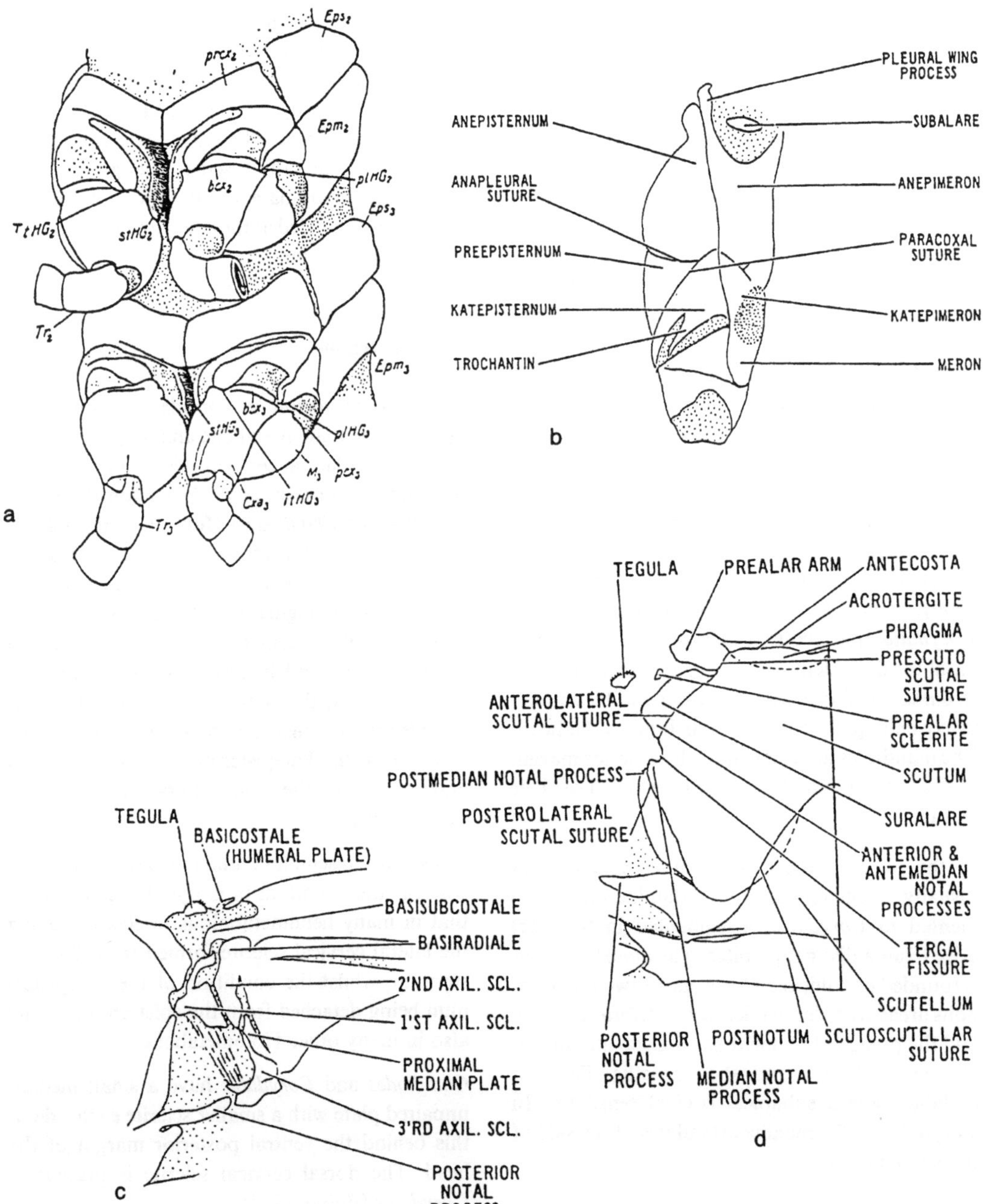


Fig. 31. Thorax of adult *Sialis* (a, b) and *Corydalus* (c, d): a, pterothorax, lateroventral; b, mesothorax, lateral; c, forewing base, dorsal; d, mesonotum, dorsal. Eps, episternum; Epm, epimeron; prcx, precoxal bridge (laterosternite); bcx, basicosta; Cxa, anterior part of true coxa; plHG, lateral articular, head of leg; TFHG, trochantinal articulatory head of leg; stHG, sternal articulation of leg; M, meron; Tr, trochanter; 2, 3, segments; membranes shaded (a, after ***Weber***, from ***Berland & Grassé*** 1951; b, c, d, ***Matsuda*** 1970).

The episternum and basicoxite are small, and the small furcasternum has relatively large lateral furcal pits. However, the morphological identity of the putative basisternum is confused, as discussed by ***Kelsey*** (1954). It is membranously separated anteriorly from the transverse jugulare (see 'cervix') and the cervical sclerites would normally be expected to articulate with the prothoracic episternum. The term 'basisternum' is probably better applied solely to the post-jugulare area, following ***Kelsey***. The precoxal bridge (precoxale), the ventral extension of the epimeral area lying anterior to the coxal cavities and which would normally join the basisternum, is indicated by ***Crampton*** (1926) and ***Kelsey*** (1954) (*Corydalus*) as crossing the midline, and thus, separating the basisternum and furcasternum. This is not shown for *Chauliodes* by ***Maki*** (1936), where the basisternum abuts the furcasternum. The precoxale is likely to be of basisternal origin in *Corydalus*, so that the presumed forward shift of the apparent basisternum is accompanied by the presence of this sclerite. A small spina is present, situated considerably beyond the furcasternum. The mesothorax and metathorax are typically wider than long. The mesonotum and metanotum are generally rather similar, and the scutellum is large and distinct in both. The antescutal suture dorsally delimits the narrow acrotergite ('precosta': ***Maki*** 1936). The prescutum is not clearly separated from the scutum in adults. In *Chauliodes* and *Corydalus* a small prealar sclerite is posterolateral to the prealar arm ('median prealar sclerite' of ***Maki***, 'laterophragma' of ***Kelsey***) which is the prescutal extension into the intersegmental membrane. The lateral parapsidal suture is absent in *Sialis* (***Czihak*** 1953) and *Corydalus* (***Kelsey*** 1957). The posterolateral scutal suture arises from the lateral scutal margin just posterior to the median notal process and reaches the base of the posterior notal wing process in *Corydalus* and *Chauliodes*. The scutoscutellar groove is very distinct. In both these genera, the postscutum is separated from the mesoscutellum by a conjunctival membrane and meets the dorsal extensions of the epimeron of each side to form a postalar bridge, and extends forwards under the wing to form a postnotal catch. The mesopostscutum abuts directly onto the metanotum, in which the prescutum has a slight median anterior membranous furrow.

Three axillary sclerites are distinctly associated with the forewing, but four with the hindwing. The tegula is well defined in both wings. The first axillary sclerite is broad. ***Kelsey*** considered it to be tergal in origin, and it articulates the wing with the tergum. The second axillary sclerite is considered both tergal and pleural in origin. Dorsally, it appears as a rectangular plate and ventrally forms a small process which articulates with a hooklike structure on the pleural wing process, aiding coordination of the wing and thorax movements in flight. It encompasses the bases of the subcostal and radial veins. A median plate, between the second and third axillary sclerites, is a thickened part of the basal wing membrane and is divided into a proximal part (articulating with the second and third axillary sclerites) and a distal part (subtending the bases of the median and cubital veins). The third axillary sclerite, the only one bearing muscles, is also believed to be tergal and pleural and extends between the median plate and the posterior notal wing process (forewing) and the fourth axillary sclerite (hindwing). The anterior portion therefore articulates directly with the median plate, and the posterior part subtends the anal veins. The metathoracic fourth axillary sclerite, presumed to be tergal in origin, is a separated part of the posterior notal wing process which permits greater flexibility of the extensive anal area of the hindwing.

In *Sialis*, the precoxal suture extends into the epimeron, dividing it into a broad dorsal anepimeron and a small ventral katepimeron (***Weber*** 1928, ***Röber*** 1941). Generally, the adult pleuron consists of the usual episternum (anterior) and epimeron (posterior) separated by a well defined median pleural suture which dorsally forms the pleural wing process. The epimeron is broad, has a deep membranous dorsal cleft containing the subalar sclerite and is dorsally extended to meet the lateral extension of the postnotum at the posterior. It is separated from the meron ventrally by a thickened postcoxal bridge. The precoxal bridge is formed by a ventral extension of the episternum, and a trochantin is present between the precoxal bridge and the coxa. It articulates with the anterior region of the coxa.

The ventral region of the thorax is difficult to interpret unambiguously, as discussed by ***Kelsey*** (1957), because the two sternal sclerites are greatly reduced and the ventral side is derived largely from pleural elements. The well-defined median groove (sometimes, 'median furrow') involves most, or all, of the basisternum, and there is no presternal plate in the *Corydalus* mesothorax. The proportion of sternal and subcoxal elements involved in a particular median groove can only be estimated, but the metathorax of *Corydalus* has a distinct remnant of the basisternum anterior to the median groove. The furcal pits are very close together because of the elevation of the furca. The ventral coxal articulations are formed from a portion of the laterosternites. The mesothoracic spinasternum is membranously separated from the poststernum and fused with the metathoracic presternum. There is no externally evident metathoracic spinasternum.

*Legs*. The legs are not remarkable, and there are few modifications of any significance or diagnostic value. The coxa is conical, the trochanter small and cylindrical, the femur and tibia both elongate and cylindrical, and the tarsus invariably 5-segmented. Two small apical tibial spurs are present, although one may be reduced (as in *Chloronia*: ***Penny & Flint*** 1982). The basal and apical tarsomeres are each longer than any of the intermediate three, and the fourth is expanded ventrally and bilobed in Sialidae and simple in Corydalidae. The paired pretarsal claws are shallowly curved and simple in most species, but can have a small subapical tooth in Chauliodinae. A pulvillus is absent, but an equivalent function may be performed by the expanded fourth tarsal segment of Sialidae.

An independently movable tibial flexor sclerite has been noted in *Corydalus*, as in some Coleoptera, Neuroptera, Hymenoptera and Heteroptera (***Furth & Suzuki*** 1990).

*Wings*. All species of Megaloptera are macropterous, with the possible exception of one anomalous South African *Platychauliodes*. ***Barnard*** (1940) recorded a single (unidentified) specimen which was micropterous, and noted that this might be an aberration. The four winglets were alike, and about as long as one third the width of the corresponding thoracic segment; each had two anterior veinlike thickenings (Fig. 32). The two pairs of wings are both normally elongate. The forewing is narrower than the hindwing, particularly in Corydalidae, as the hindwing anal field is expanded. The wing membrane of Sialidae is almost always somewhat fumose, sometimes with the basal region darker than the apex, occasionally with the whole wing almost black. Wings of Corydalidae may be more elaborately patterned. In Corydalinae the wings range from being nearly hyaline, pale yellowish, through greyish or brownish with darker markings, to more contrasted mosaics of white maculation on black ground colour. Chauliodinae also vary in pattern, and some genera are characteristically spotted; others have hyaline banding across fuscous wings.

The wings are held rooflike over the body when the insects are at rest, and the steepness of the 'roof' reflects wing width. In broad-winged forms such as *Chloronia*, for example, the arch is very shallow. *Taeniochauliodes*, resting on bushes overhanging streams, may have the wings wrapped

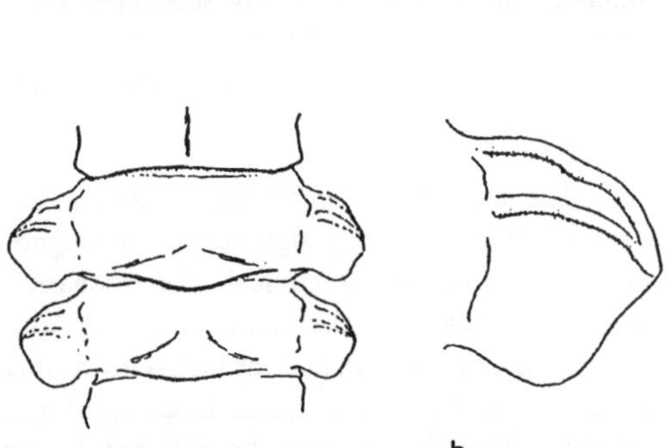


Fig. 32. *Platychauliodes* sp., microptery: a, thoracic segments with winglets; b, one winglet enlarged (***Barnard*** 1940).

around the twigs (***Barnard*** 1931). Sialid wings have numerous microtrichia on the upper membrane surface (forewing) or lower surface (hindwing). ***Glorioso*** (1981) notes that wings of Corydalinae are densely covered with microtrichia, but those of many Chauliodinae are bare. *Archichauliodes* has a few macrotrichia in some of the hindwing costal cells. A jugal lobe is more or less developed in the forewing, often with a well-developed jugal vein.

Venation (Figs 11, 12, 16, 17) is relatively complete, and all veins are well sclerotised, usually well defined and clearly differentiated from the membrane. Details are of considerable taxonomic value, and may be diagnostic at both generic and specific levels. Most veins are glabrous, though the costa is usually densely trichiate, and microtrichia may also occur on the radial and medial veins and in the anal region.

The costal cell is usually narrow, but broad in *Sialis*, and traversed by crossveins. They are most usually simple, but sporadically forked in many genera and form a reticulate pattern in male *Acanthacorydalis*. Their number ranges from less than a dozen (Sialidae) to more than 60 (larger Corydalidae). The pterostigma is usually not defined in Corydalidae, although costal crossveins may be dense towards the wing apex. Veins Sc and R1 run closely parallel for much of the wing length and fuse distally with R, reaching the wing margin well before the wing apex. Vein Rs commonly arises at about a quarter of the wing length and is pectinate with 4–16 branches in Corydalidae and fewer in Sialidae; the posterior branch is forked in most Corydalinae (except *Corydalus* and *Chloroniella*). Extensive apical 'twigging' is absent, but more anterior branches of Rs may be forked: in Sialidae, vein R2 + 3 may have 2,3,4 or 5 branches in the forewing, and up to 4 branches in the hindwing; Chauliodinae may have both R2 and R3 forked (forewing) and various radial branching patterns in the hindwing are characteristic of particular genera. A basal Sc crossvein is sometimes present, and there are usually few (most commonly, 3) R1 crossveins linking with the radial sector. Some Corydalinae have 4 or more such crossveins. Some authors (eg. ***Aspöck et al.*** 1980, ***Riek*** 1970) consider that MA and Rs are basally contiguous so that the posterior branch of the apparent Rs is, in fact, MA. The medial vein is basally close to the radius. MA forks beyond the origin of Rs (Sialidae, Corydalinae), and the branching of both MA and MP is, again, often generically diagnostic in Corydalidae. The cubital vein arises basally and diverges from M. In Corydalinae, Cu1 has 3 or more branches in all genera except *Chloroniella* (forewing and hindwing) and *Protohermes* (hindwing only). In Chauliodinae vein Cu1 is usually forked, and vein Cu2 is simple in Corydalidae.

Three anal veins are present in Corydalidae; 3A is absent in Sialidae. The first two anal veins are usually forked, but not in the hindwing of *Corydalus* and *Acanthacorydalis*. Patterns of crossveins throughout the wings are often diagnostic. In general, the venation is rather 'open', and only in *Acanthacorydalis* are there commonly numerous supernumary crossveins. Crossvein addition has been a general trend in Corydalinae (***Glorioso*** 1981).

The base of the anterior veins of the wing of female *Sialis* contain possible scent glands which have been suggested to be used in attracting males (***Geigy & DuBois*** 1935).

## Abdomen

The abdomen of Megaloptera is relatively soft and flexible. Anteriorly it is cylindrical. The first segment is shorter than more posterior non-genital segments, and the tergum is membranously divided into the main region and what ***Maki*** (1936) referred to as the posterior plate or precosta of the following segment. The first tergum is then separated from the second by a broad shallow thickened infolding (or antecosta of segment 2). Laterally, the tergal border is indistinct. The first sternum is also transverse and also membranously divided into anterior and posterior plates, the latter narrow. As in the tergum, the lateral margin is indistinct in *Chauliodes* and there is a posterior sternal infolding as the second sternal antecosta. Lateral margins of both sclerites are more distinct in some other taxa. Other pregenital segments are weakly sclerotised and the quadrate tergum and sternum are similar to the first segment. One exception to this is the 'sternal pouch' of females of *Platyneuromus, Chloronia* (Fig. 34) and some *Corydalus* (***Glorioso*** 1981). This structure, on the sixth sternite, is semicircular and membranous, and opens into the intersegmental membrane; its function is unknown and until ***Glorioso***'s survey had been reported only in *Chloronia* (***van der Weele*** 1910). The pleural region is very weakly sclerotised, so that the abdomen often appears somewhat flattened dorsoventrally. In dry-preserved specimens the abdomen usually 'wrinkles' or contracts to a considerable extent. Segment VIII is sometimes shorter than the immediately preceding segments and the more posterior segments are modified extensively to form the external genitalia. Some homologies in this region are by no means clear and understanding can be aided by reference to more general accounts of the insect abdomen (such as ***Matsuda*** 1976) and studies of larval development (see later). ***Achtelig***'s (1975) interpretation of the anterior abdominal segments of *Sialis* is shown in Fig. 33.

## Female external genitalia

Structural modifications predominantly involve segment VIII (Sialidae) or IX (Corydalidae) and more posterior regions. The eighth sternite is

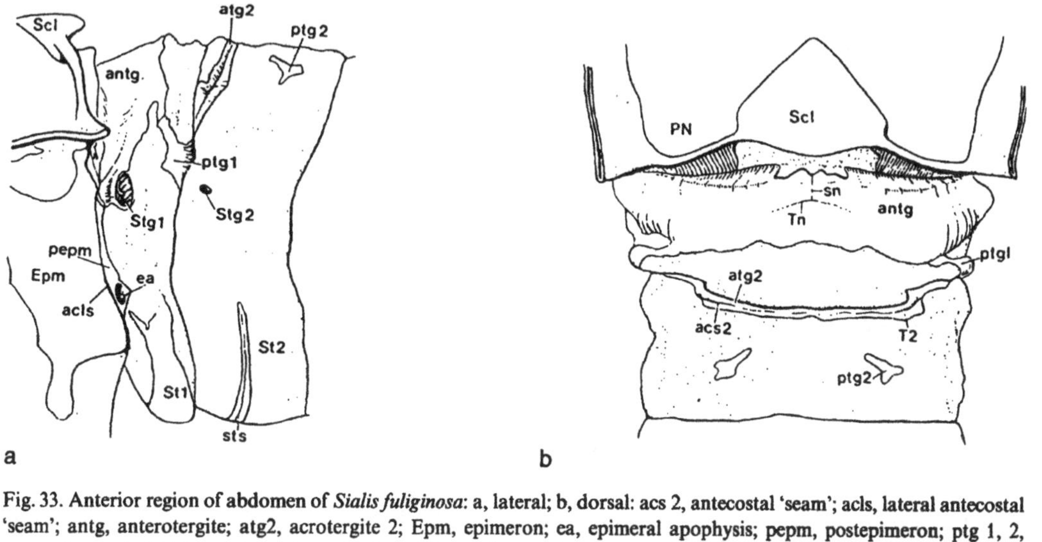


Fig. 33. Anterior region of abdomen of *Sialis fuliginosa*: a, lateral; b, dorsal: acs 2, antecostal 'seam'; acls, lateral antecostal 'seam'; antg, anterotergite; atg2, acrotergite 2; Epm, epimeron; ea, epimeral apophysis; pepm, postepimeron; ptg 1, 2, posterotergite 1, 2; PN, metathoracic postnotum; Scl, scutellum; St1, 2, sternites 1, 2; Stg 1, 2, spiracle 1, 2; sts, sternal 'seam' (***Achtelig*** 1975).

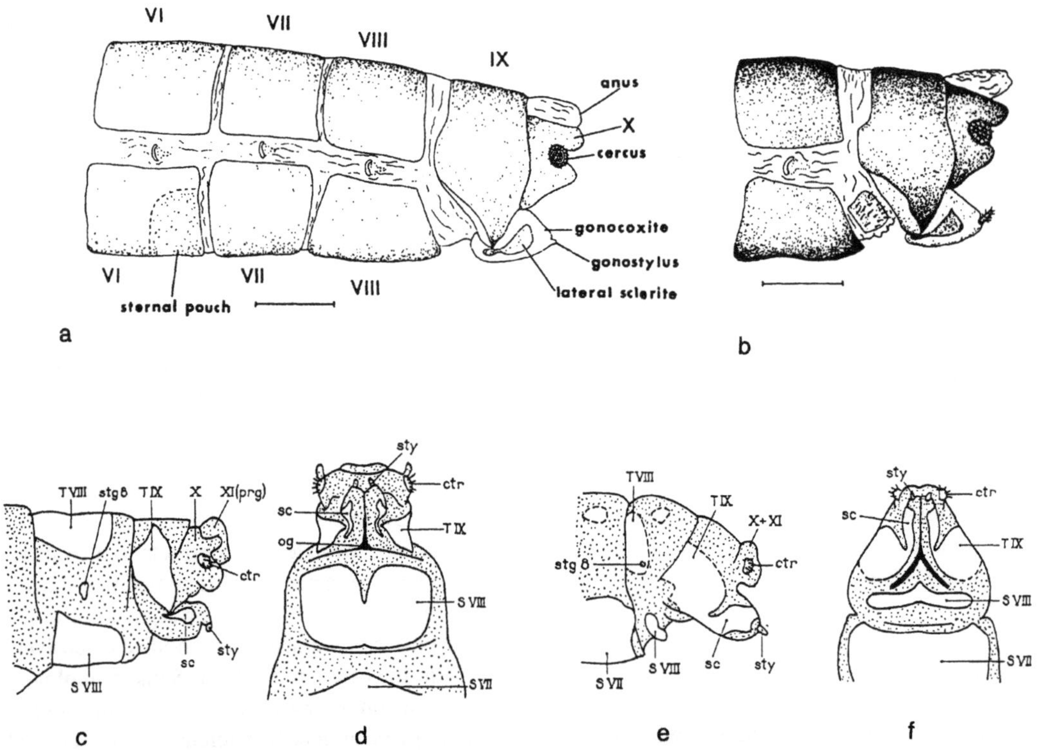


Fig. 34. Female abdomen of a, *Chloronia; b, Neurhermes*, both lateral aspect; scale line is 1 mm (***Glorioso*** 1981); c–f, lateral (c, e) and ventral (d, f) aspects of *Corydalus* (c, d) and *Sialis* (e, f): ctr, cercal callus; og, genital opening; prg, proctiger; sc, lateral sclerite; stg, spiracle; sty, style (from ***Bitsch*** 1979, after ***Mickoleit*** 1973).

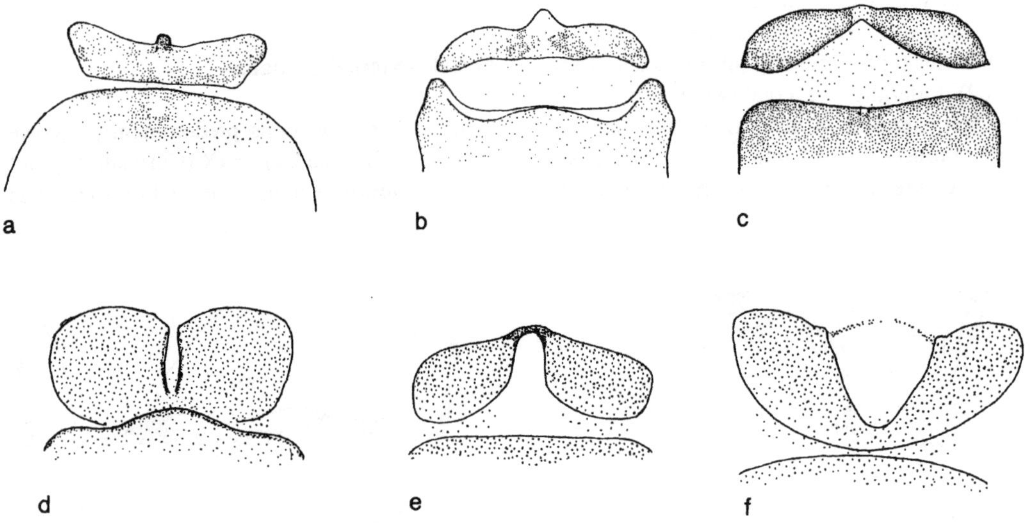


Fig. 35. Modification of abdominal sternite VIII in female *Sialis*, ventral aspect: a, *S. lutaria*; b, *fuliginosa*; c, *nigripes*; d, *nevadensis*; e, *rotunda*; f, *velata*. (a–c, after ***Aspöck*** et al. 1980; d–f, after ***Ross*** 1937).

relatively unmodified in Corydalidae (Fig. 34). but is more extensively changed in Sialidae (Fig. 35) where it is normally a narrow band with a straight caudal margin. It may be constricted, and very narrow, in the midline. Medially there is usually a triangular sclerite posteriorly, which is more or less distinctly separated from the rest of the sternite and which may protrude posteriorly to give a pointed central margin. In some species, such as *'Protosialis' bimaculata* and *mexicana*, the sternite is more extremely reduced to a small triangular sclerite. In *Austrosialis*, sternite VIII is more elongate and deeply incised posteriorly in the midline. *Protosialis chilensis* appears to be very distinctive in having this sternite well sclerotised and the ninth sternite divided into a pair of elongate central sclerites and two lateral plates below tergite IX.

Generally, the ventral area immediately behind sternite VIII is membranous, but there are two weakly sclerotised plates in *Neurhermes* (Fig. 34). They are unique to this genus. Tergite IX is moderately to well sclerotised, elongated dorsoventrally and divided by a membranous area in the dorsal midline. Tergite X is dorsally divided to form the ectoprocts, which are oval and often very small in Sialidae but may be deeply bilobed in Corydalidae, and typically each bears a large circular cercal callus with a number of trichobothria.

The lateral gonapophyses, or gonocoxites, form the ovipositor. That of each side arises near the ventral angle (which is sometimes strongly tapered) of tergite IX, and usually has a small conical subapical style, or gonostylus. This is articulated with the gonocoxite in *Neurhermes* and *Protohermes* but fused with it in other taxa. In Corydalidae, particularly, a strengthening lateral sclerite is present. This is often heavily sclerotised, though not in *Chloronia* (***Glorioso*** 1981). The genital opening is in the membrane near the base of the gonocoxites, and the form of the bursa copulatrix varies. It can, for examples, be membranous (*Corydalus*) or muscular and/or tubular (*Neoneuromus, Neuromus*). This seems to have evolved independently in different lineages (***Glorioso*** 1981).

## Male external genitalia

Elements of segments IX and X comprise the male external genitalia, but there is still confusion over the homologies of particular structural components, as in other neuropteroids, and a number of descriptive terms have been employed without precise understanding of their morphological derivation. A detailed glossary, with discussion is provided by ***Aspöck et al.*** (1980). In Sialidae, sternite IX varies considerably in form (Fig. 36): it may be simple, quadrate, elongate and oval in *Sialis*; extended with a central apical point (*Stenosialis*) or bifurcate (*Austrosialis*); or very narrow so that it is less than half the width of the abdomen in *Protosialis chilensis*. In Corydalidae (Fig. 37) it is more frequently quadrate, and wider than long. Again, variations occur in particular taxa so that, for examples, *Acanthacorydalis* has posterolateral lobes and *Orohermes* has the posterolateral angles produced to form elongate projections. In some Chauliodinae, sternite IX is small, reflecting the large ventral extension of tergite IX enclosing part of the ventral surface of the body (e.g. *Archichauliodes, Neohermes*). The tergite, which is usually more heavily sclerotised than the preceding segments, is also extended ventrally in Sialidae – this extension is more evident at the anterior end of the tergite. In some Corydalinae, it has an arched or inverted V-shaped basal internal inflexion. The tergite is much wider than long in *Neurhermes*, but is otherwise generally nearly quadrate. Tergite X is sometimes divided medially in Sialidae and, in its most simple form the ectoprocts form a ring around the anus (*Sialis californica*: Fig. 36). The tergite may be modified

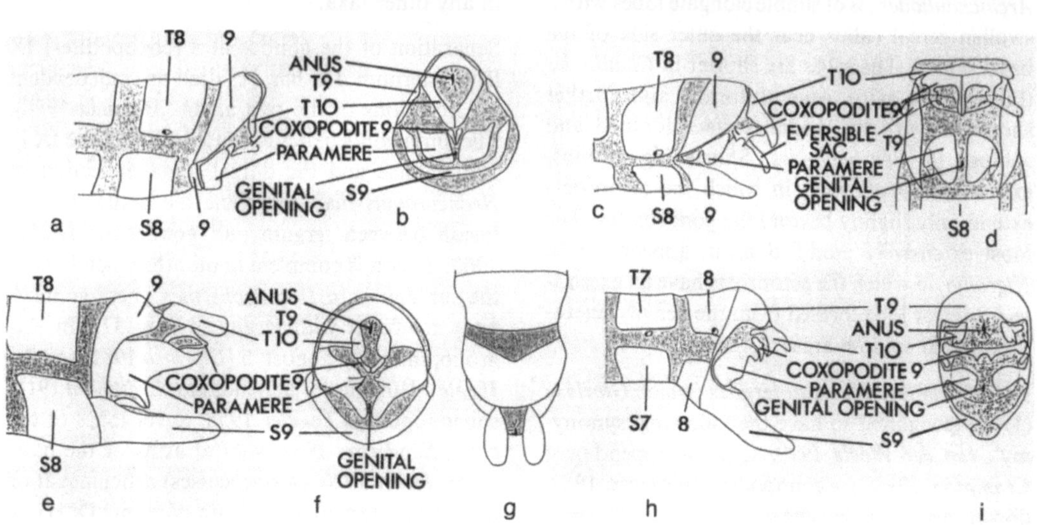


Fig. 36. Male genitalic structures of Sialidae: a, b, *Sialis californica*; c, d, *S. lutaria*; e, f, g, *S. nevadensis*; h, i, *S. chilensis*(a,c, e, h, lateral; b, f, i, caudal; d, ventral, sternite 9 removed; g, dorsal; schematic to indicate terminology of ***Acker*** 1960) (***Acker*** 1960).

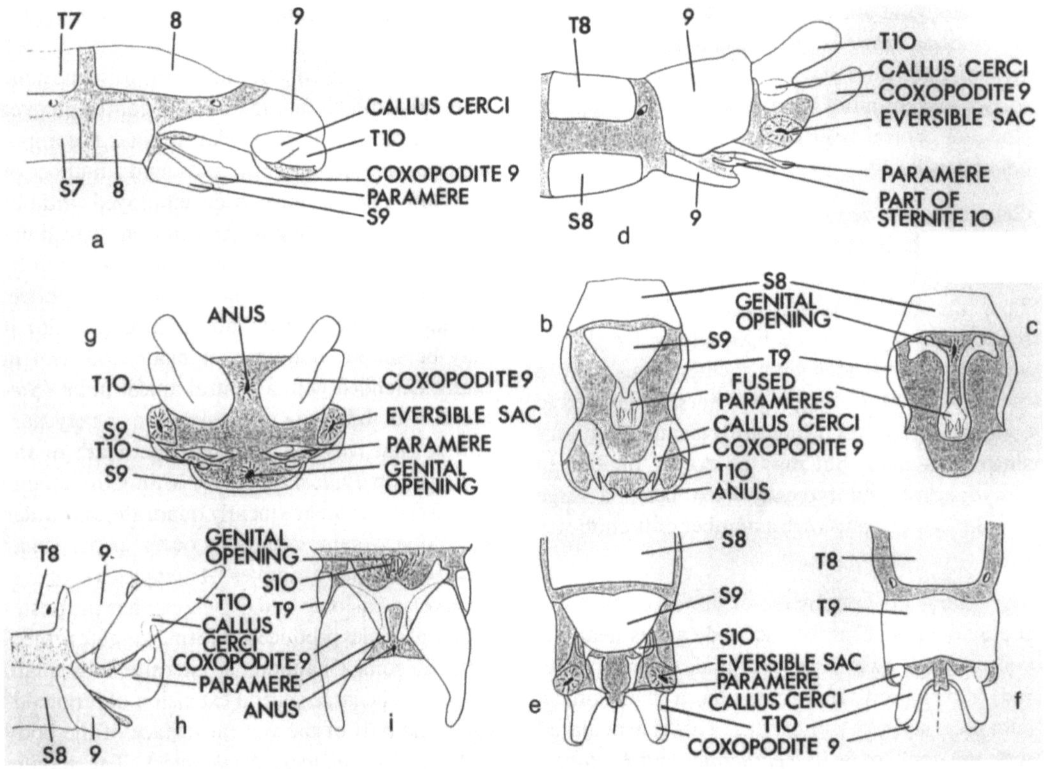


Fig. 37. Male genitalic structures of Corydalidae, to indicate terminology: a, b, c, *Nigronia fasciatus*; d, e, f, g, *Dysmicohermes ingens*; h, i, *Chauliodes pectinicornis* (a, d, h, lateral; b, e, ventral; c, i, ventral; sternite 9 removed; f, dorsal; g, caudal) (***Acker*** 1960).

in various ways, such as being divided and extended to form swollen digitate lobes (*S. nevadensis*: Fig. 36) or with separate ectoprocts reduced to small hooked sclerites, as in *P. chilensis* (Fig. 36). In Chauliodinae the tenth tergites are amongst the most variable male genitalic features. The presumed generalised form, as represented by *Archichauliodes*, is of simple elongate lobes with a swollen cercal callus near the outer side of the basal region. The lobes are shorter in *Chauliodes* (but of the same general form), and further shortened with the medial surface toothed and spinose in *Neochauliodes*. Shortening is most extreme in *Orohermes*, in which the ectoprocts extend only slightly beyond the gonocoxites. The most extensively modified genus appears to be *Nigronia*, in which the ectoprocts have an extensive posterior lobe formed from the heavily sclerotised mesal spinose area.

In Corydalinae the tenth tergites (which ***Glorioso*** (1981) recognised to have the following 'synonymy': ***van der Weele*** 1910: superior appendages; ***Crampton*** 1918: surgonopods; ***Snodgrass*** 1957: dorsal arms of basimere; ***Tjeder*** 1970: anoprocesses) are generally digitate, more rarely clavate, and range in length from about equal to that of tergite IX to 2–3 times as long. They are flattened in some *Neuromus*, and may be biramous (*Neohermes*, some *Protohermes*). Cerci are small and hemispherical and separated from the tergites to lie between them and the gonostyli. In *Neoneuromus* there is a small suranal sclerite behind tergite X, which may represent the eleventh tergite (***Acker*** 1960), but this is not evident in any other taxa.

Separation of the gonocoxites (coxopodites) IX from sternum IX has resulted in independent modifications of the two areas (***Matsuda*** 1976). The ninth gonocoxites are fused with tergite IX in Corydalinae and the dorsal tergal inflection of *Neoneuromus* may represent the actual line of fusion between tergum and gonocoxite (***Acker*** 1960; fusion is complete in all other members of the subfamily. In *Dysmicohermes*, these gonocoxites are fused with tergite X (Fig. 37). In this group, the ninth gonostyli (***Glorioso*** 1981; ***van der Weele*** 1910: inferior appendages; ***Crampton*** 1918: subgonopods; ***Chandler*** 1956: lower lobes of the cerci; ***Snodgrass*** 1957: ventral arms of the basimere; ***Tjeder*** 1970: catoprocesses) articulate at or near the posterolateral margin of tergite IX. They can be clavate (*Acanthacorydalis*) and densely setose, orientated posteromedially and may end in a small sclerotised 'claw'. Unguiform styli occur in other genera, such as *Neuromus* and *Protohermes*, and are directed medially. They may have a glabrous tip or be more extensively glabrous. Other orientations of the gonostyli occur in Corydalinae, but gonostyli IX are absent in Chauliodinae – a characteristic of that subfamily which renders their genitalia relatively simple, so that the fused gonocoxites X (***van der Weele*** 1910; ***Crampton*** 1918: penis; ***Tjeder*** 1970: parameres; ***Chandler***, 1956: aedeagus; ***Acker*** 1960: parameres) form a single plate dorsal or posterior to sternite IX and between tergites X. It is apically notched in *Chauliodes* (Fig. 37) and some others, and is most developed in *Orohermes* in which the gonostyli form long processes extending almost to the apex of tergites X.

The gonocoxites X normally form a transverse band, but there is a median projection in *Neoneuromus*. The styli are commonly digitate and setose and are posterolateral. They may be broadly fused (some *Neuromus*) or shortened and papilliform (*Corydalus, Chloronia*). A pair of genital papillae (***Glorioso*** 1981: utriculi of ***Tjeder*** 1970) are present in some Corydalinae as structures on each side of the genital opening. These possibly represent apophyses of sternite X (***Acker*** 1960) or may originate from sternite IX. They appear to represent a primitive condition in Corydalinae and have been lost in *Neoneuromus, Chloronia* and others.

The 'parameres' (coxopodites IX, gonocoxites) remain separate in Sialidae, and may form broad plates. They are occasionally elongate with an apical hook (*S. nevadensis*: Fig. 36) or may be dentate near the apex (*S. nigripes*). Sialidae also have a more dorsal genitalic complex which resembles that of many Planipennia. This has been variously claimed to be derived from coxopodite IX, the styli or sternite X, but more descriptively comprises a broad plate, the 'gonarcus', which most simply is a pair of triangular plates juxtaposed along the midline. The lateral gonarcus margins can be recurved anteromedially to form a pair of lateral fossae which support membranous eversible sacs. In some neotropical taxa formerly placed in *Protosialis*, the lateral margins of the gonarcus are extended to form dorsolateral 'wings'. The gonarcus has a median dorsal apical process, the mediuncus, which sometimes has paired hooks and can be considerably elongated in some North American taxa. In contrast, many Palaearctic *Sialis* have a small mediuncus with a small single median posterior hook.

## Sexual dimorphism

Several features noted above relate to various forms of sexual dimorphism in Megaloptera. Other than genitalic differences, sexual dimorphism is sporadic, and in most species the two sexes are closely similar. In some taxa, the difference is merely one of size as females tend to be larger than males (in many *Sialis*, for example). Despite size variations between different streams, male *Protohermes grandis* are significantly smaller than females (***Hayashi*** 1988a). More conspicuous dimorphisms occur, predominantly as modifications of the head and its appendages

i) Mandibles of *Corydalus* and *Acanthacorydalis* males may be greatly enlarged.
ii) Mandibles of male *Sialis* are more strongly toothed or elaborated than the relatively smooth mandibles of the corresponding females.
iii) *Protosialis* male antennae are more densely hirsute than those of females. Likewise, male antennae of *Neohermes* have whorls of erect hairs on most segments.
iv) Antennal structural dimorphism occurs in some genera or species of Chauliodinae, with the male antenna being the more elaborate. As examples: *Neochauliodes* – male pectinate, female subserrate; *Parachauliodes* – male serrate, female subserrate.
v) *Sialis* males have labral sense organs (sensilla

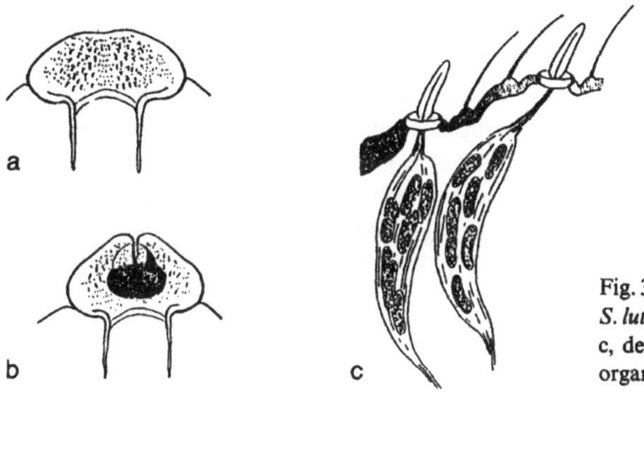


Fig. 38. Labral sensilla of male *Sialis*; sexual dimorphism in *S. lutaria*. Ventral aspect of (a) female and (b) male labrum; c, detail of two sensilla basiconica from the male labral organ (***Geigy & DuBois*** 1935).

Fig. 39. Sexual dichromism in *Protochauliodes kirramae*. A, male; B, female (forewing lengths: A, 35 mm; B, 42 mm).

basiconica) (Fig. 38) (***DuBois & Geigy*** 1935), possibly used to detect scents from glands on the wing bases of females.

vi) The postocular flange in *Platyneuromus* males is larger than in the female (Fig. 14). Here, however (as with *Corydalus* mandibles), it may be relatively more enlarged in large individuals so that there is size-related variation as well as sexual dimorphism. In *P. soror*, a very large flange on a particularly large individual induced ***Navás*** (1925) to describe the insect as a new species, genus and tribe (*Doeringia christel*: Doeringiini)!

vii) Very rarely, other (postcranial) factors may differ between sexes. The male wings of the Australian *Protochauliodes kirramae* are almost completely blackish brown, but those of the female are pale, so that the species is sexually dichromic (Fig. 39).

Little other (non-sexual) variation has been recorded. However, ***Hayashi*** (1990 a) noted a probable instance of insular dwarfism in *Protohermes*. Earlier, he noted (***Hayashi*** 1988 a) that body size of *P. grandis* tended to reflect prey availability, so that individuals in populations from streams poor in large prey were smaller than when prey was abundant. *P. costalis* from Taiwan is larger than a closely related species from Iriomote and Ishigaki islands some 200 km to the east, and low prey availability on the islands may be related to this. Geographical variation in intensity of wing pigmentation is evident in *Neochauliodes sinensis* (***Hayashi*** 1990 b) (Fig. 40). Thoracic stripe patterns of *P. grandis* are variable (***Hayashi*** 1988 b), possibly polymorphic.

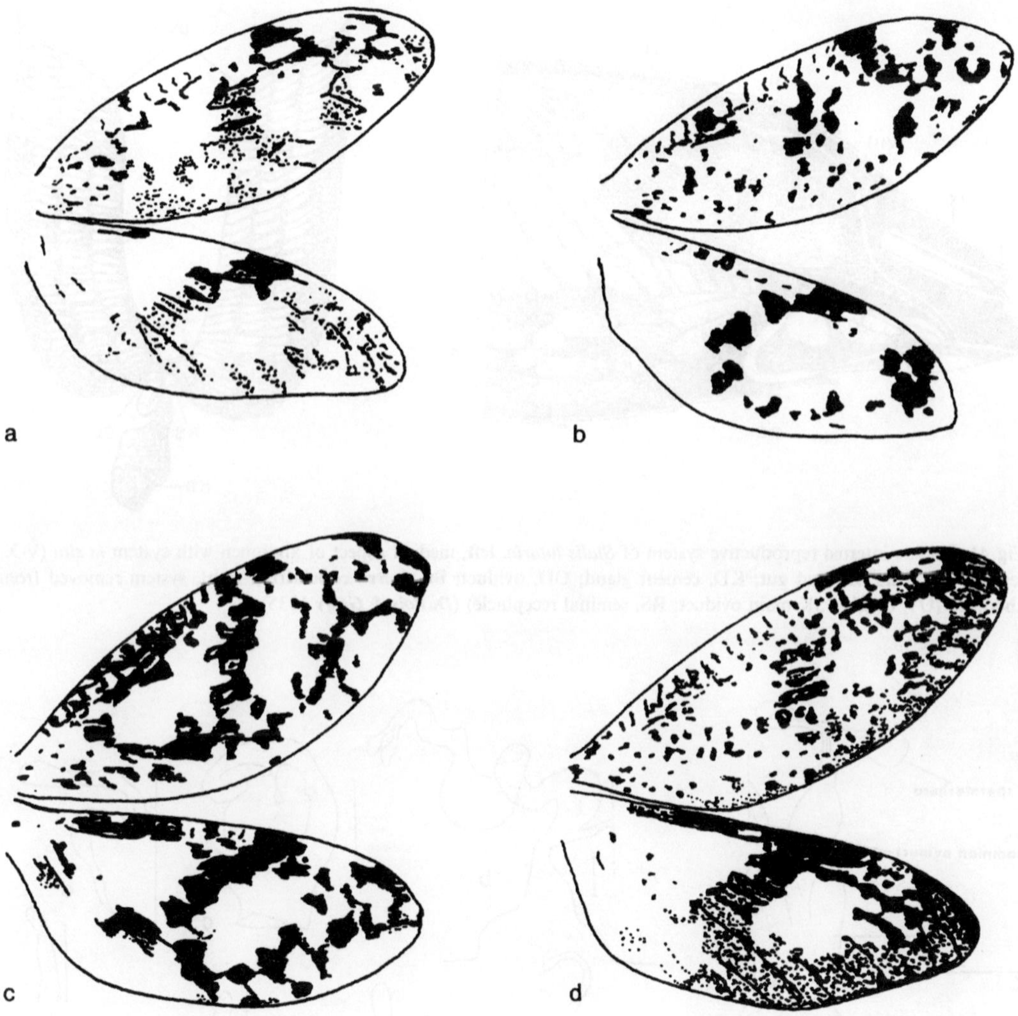


Fig. 40. Geographical variation in wing pattern of *Neochauliodes sinensis*: a, Tsushima Is; b, Amami-Oshima Is; c, Iriomote Is; d, Taiwan (based on ***Hayashi*** 1990b).

## Anatomy and Physiology

### Female internal reproductive system

The anatomy of the female internal reproductive system of *Sialis* was studied by ***Gross*** (1903) and ***DuBois & Geigy*** (1935), and the latter depict the appearance of the whole system (Fig. 41). A large basal gland was interpreted as a 'cement gland' and opens immediately behind the opening of the bursa copulatrix into which the common oviduct opens. In Corydalinae the spermathecal duct opens into the bursa or may be continuous with it. There may be a more discrete duct with a T-shaped transverse apex (so that there are, essentially, two spermathecae: *Corydalus cornutus*, Fig. 42), or a single spermatheca on a more continuous duct (*Neuromus*, Fig. 42). *Neoneuromus* has two spermathecae on the continuous duct (Fig. 42) and ***Glorioso*** (1981) considered a single spermatheca and the duct discrete from the bursa to be apotypic features. Accessory glands in Corydalinae may be present or absent, short or long, and sigmoid or linear (Fig. 42). They are particularly long in the *Protohermes-Neurhermes* group, and are lacking in all New World genera.

The paired ovaries (Fig. 41) are large, and consist of numerous ovarioles. Each has around 100 ovarioles in *Sialis*, where they are arranged in four parallel rows, and 145–150 ovarioles in *Protohermes*. The form of the ovarioles differs between the two families. In *Sialis fuliginosa*, ***Gross*** (1903) claimed that they were telotrophic, an opinion doubted by ***Bonhag*** (1958) who pointed out that,

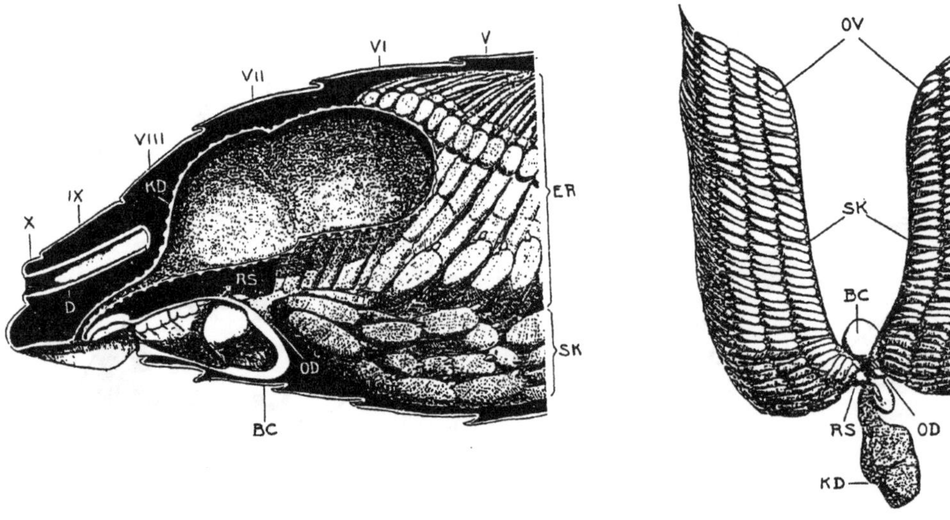


Fig. 41. Female internal reproductive system of *Sialis lutaria*: left, median aspect of abdomen with system *in situ* (V-X, segment numbers; D, hind gut; KD, cement gland; OD, oviduct; BC, bursa copulatrix); right, system removed from abdomen (OV, ovaries; SK main oviduct; RS, seminal receptacle) (***DuBois & Geigy*** 1935).

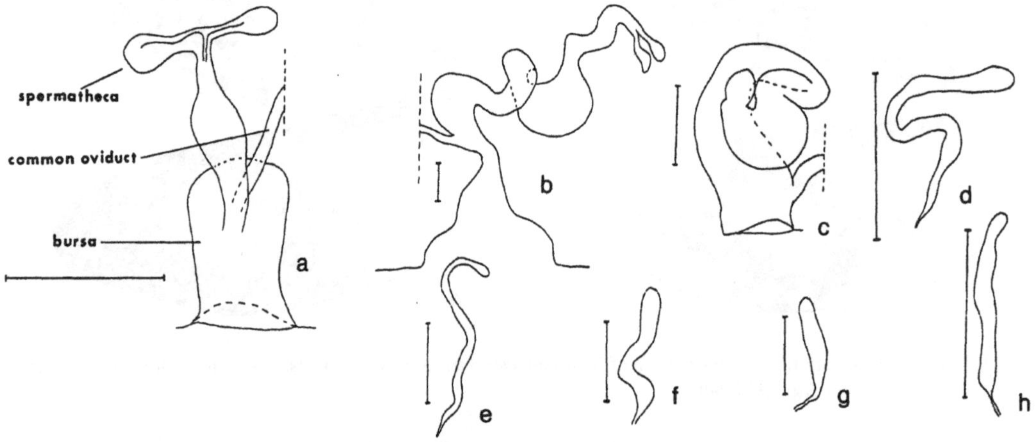


Fig. 42. Spermathecal form, bursa (a–c) and accessory glands (d–h) in female Corydalidae: a, *Corydalus cornutus*; b, h, *Neoneuromus*; c, g, *Neuromus*; d, *Neurhermes*; e, *Protohermes*; f, *Acanthacorydalis*; scale line is 1 mm (***Glorioso*** 1981).

with knowledge then available, if this was so *Sialis* would be the only known non-polytrophic genus in Neuroptera. Gross described the large terminal chamber but was unable to demonstrate nutritive cords. However, *S. mitsuhashii* clearly has telotrophic ovarioles (***Matsuzaki & Ando*** 1977), with the nurse cells located in the germarium. The germarium itself consists of a peripheral cell layer and an extensive syncytial nutritive tissue. Interstitial cells occur, around the peripheral cell layer and in the spaces between the cells, as in Coleoptera (***Matsuzaki & Ando*** 1977, ***Büning*** 1979), though the ovarioles themselves differ significantly from those of beetles. No differences were found from *S. mitsuhashii* for ovarioles of *S. flavilatera* (a synonym of *S. lutaria*) by ***Büning*** (1979). The ovarioles of *Sialis* are of the same basic kind as those of *Raphidia* (***Büning*** 1980), adding support to these being sister groups. The terminal filament in *Sialis* is not separated from the terminal chamber by a transverse septum but continues into the somatic cells of the inner sheath. The ovariole is surrounded by a tunica propria and a poorly developed outer envelope. The inner sheath (see Fig. 43) continues into prefollicular tissue which surrounds each developing oocyte, and the somatic tissues are similar to those of other meroistic ovarioles in structure (***King & Büning*** 1985).

In contrast, *Protohermes grandis* (the sole member of the Corydalidae for which published informa-

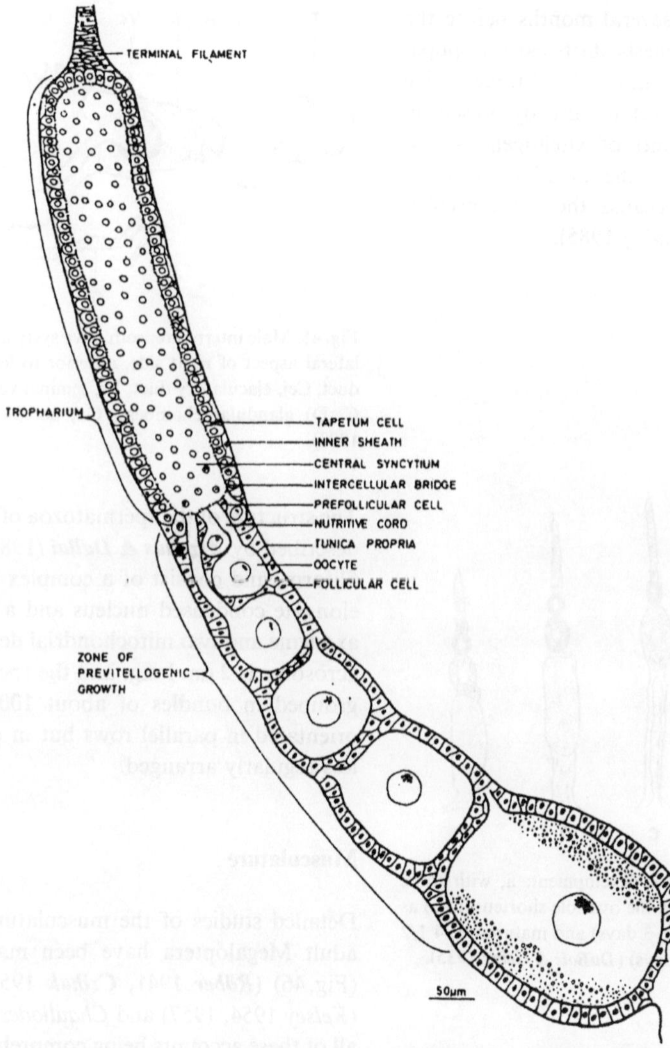


**Fig. 43. Ovariole structure of *Sialis lutaria*: telotrophic meroistic ovariole showing different regions. Each tapetum cell has an intercellular bridge to the central syncytium (*King & Büning* 1985).**

tion is available) has panoistic ovarioles (***Matsuzaki et al.*** 1985) with no nurse cells, so that there appears to be a basic dichotomy in this feature in the Megaloptera.

*Sialis* larvae towards the end of their growth can be sexed, both by head capsule width and body weight, and also by the gonads being easily distinguishable from about instar VII onwards (***Pritchard & Leischner*** 1973). The testes of such larvae were described as 'resembling a string of beads' on each side of the intestine, and the ovaries are striated ellipsoidal bodies.

Germ cells occur first in the gonadal tube, which develops later into the oviduct. The first-differentiated tissue is the terminal filament, and it continues into the inner sheath cells (Fig. 43) before these form a single layer of tissue and fill the whole terminal chamber with large vacuolated cells. They are displaced by germ cells migrating from the gonad tube (***King & Büning*** 1985), a process which lasts for several months in the larva. Eventually, germ cell clusters are produced as their number increases by migration and mitosis and a developmental gradient is established, whereby apical cells continue mitotic activity but basal germ cells enter meiotic prophase. The clusters remain interconnected by cytoplasmic bridges towards the central region, resulting in cellular 'rosettes'. They each contain $2^n$ cells, where n is the number of consecutive divisions, with the probable value 5, according to the model proposed by ***King & Büning***. In fully developed ovarioles cluster formation has ceased and the central syncytium (surrounded by tapetum cells) fills the terminal chamber. The definitive oocytes start to develop at the base.

Thus, in *Sialis* previtellogenesis commences during the larval stage, several months before the pupal moult. Vitellogenesis starts early in pupal life so that at the adult moult about three to five chorionated eggs (Fig. 44) are already present in each ovariole. This shift of vitellogenesis and chorion formation into the preadult stages is presumed necessary because the adult itself is short-lived (***King & Büning*** 1985).

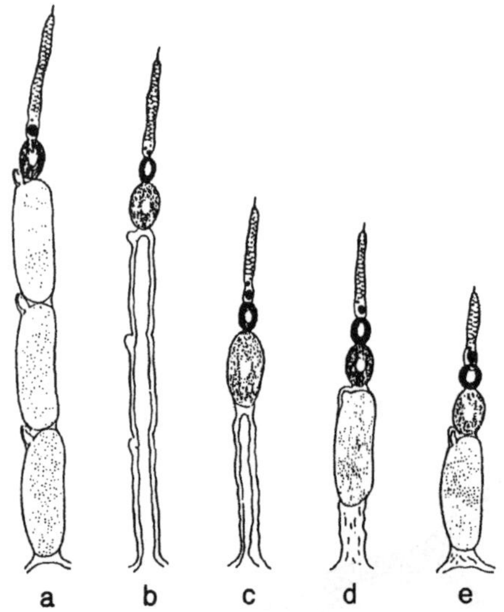


Fig. 44. Sequence of oocyte development: a, with three mature eggs. When laid (b), the ovariole shortens (c–e) as the next follicle enlarges (c, 3 days) and matures (d, 4 1/2 days) ready for laying (6 days) (***DuBois & Geigy*** 1935).

## Male internal reproductive system

Little detailed information is available on the male internal reproductive system in Megaloptera, and most authors have repeated the description of *Sialis* in ***Stitz*** (1908, see also ***Stitz*** 1926) (Fig. 45) and assumed that this is fully representative. The system had earlier been depicted by ***Loew*** (1848), who also clearly shows six major divisions of the reniform testis (as does ***Dufour***, 1841), and by ***Suckow*** (1828). There appears to be little which can be considered unusual. The vas deferens from each of the anterolateral testes is long and slender. It extends almost to the abdominal apex, then loops anteriorly to there join a large seminal vesicle comprising several large glandular lobes. There is an additional slender sinuous accessory gland near the midline, and the single efferent duct from each vesicle is short. The short common ejaculatory duct is also slender.

In *Corydalus* the testes are also well anterior in the abdomen: in the pupa they may extend as far forwards as the metathorax (***Leidy*** 1848).

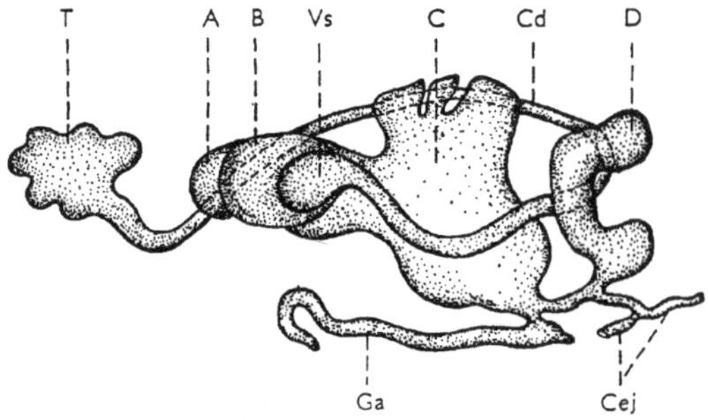


Fig. 45. Male internal reproductive system of *Sialis lutaria*, lateral aspect of right side, anterior to left: T, testis; Cd, duct; Cej, ejaculatory duct; Vs, seminal vesicle with (A, B, C, D) glandular chambers; Ga, accessory gland (***Stitz*** 1908).

The structure of the spermatozoa of *S. lutaria* was described by ***Afzelius & Dallai*** (1988). Sperm are filiform, and consist of a complex acrosome, an elongate condensed nucleus and a tail with one axoneme and two mitochondrial derivatives. The acrosome is 2 µm long, and the spermatozoa are grouped in bundles of about 1000, sometimes orientated in parallel rows but in other bundles less regularly arranged.

## Musculature

Detailed studies of the musculature of parts of adult Megaloptera have been made for *Sialis* (Fig. 46) (***Röber*** 1941, ***Czihak*** 1953), *Corydalus* (***Kelsey*** 1954, 1957) and *Chauliodes* (***Maki*** 1936), all of these accounts being comprehensively illustrated and providing lists of the muscles present and details of their positions of insertion. ***Kelsey*** (1954) regarded ***Maki***'s study as until then 'probably the most profound study of the skeleto-motor mechanism of any member of the Neuroptera,' and it is the only one to include all body regions. ***Röber*** assessed only the head and thorax, but of both larval and adult *Sialis*, ***Czihak*** the thorax only, and ***Kelsey*** dealt with these areas of larval, pupal and adult *C. cornutus* and emphasised the use of this species as a relatively primitive holometabolous insect which could well illustrate homologies between these different stages. ***Kelsey*** showed clearly, for example, that most larval cranial muscles persist through to the adult stage without changing their points of attachment during the pupal stage (***Matsuda*** 1965). Homologies of the mouthpart muscles, and their comparison with *Raphidia*, are tabulated by ***Achtelig*** (1967). ***Matsuda*** (1970) homologised the thoracic muscles of the three insects and emphasised that, in this region, some larval muscles do become lost or change their positions in development. In *Coryda-*

*lus*: (1) some muscles become lost in some segments only, and their serial homologues are retained in others; (2) muscles lost in the adult *Corydalus* may be present in adults of other neuropteroids; (3) muscles may be, or become, fused with a neighbouring muscle; (4) muscles may become, by reorganisation, adult muscles with different designations; (5) muscles may be lost in the adult as with other neuropteroids, and (6) a few muscles were regarded as 'questionable' as their fate was not ascertained. There are also a few adult muscles which develop only during the pupal stage. ***Matsuda*** (1970) cited the presence of a muscle (designated t-s? 10) connecting the anterolateral part of the pronotum and the posterolateral part of the prothoracic basisternum as noteworthy, as it agreed in position with this muscle in Odonata except there it is inserted on the apodeme of the basisternum. Muscle p-s 7 (in the cervico-prothorax region) had been found only in *Corydalus* among adult insects, and several other relatively unusual muscles occur in neuropteroids.

***Kelsey*** (1954) noted that the head muscles of *Corydalus* agree closely with the generalised 'orthopteroid scheme', as remarked also by ***Maki*** (1936) for *Chauliodes*. The second adductor muscle to the mandibles, found in *Chauliodes*, is absent in *Corydalus*. In the prothorax, *Corydalus* has several suspensory muscles for the large cervical and prothoracic intersegmental membrane areas, and these are not found in insects with more restricted membranous areas. There are also minor differences in prothoracic leg muscles between *Corydalus* and *Chauliodes*, namely (1)

Fig. 46. Muscles of thorax and abdomen of *Sialis*: a, b, right side of thorax from midline, b with major flight muscles removed to show more lateral muscles; c, d, right side of abdominal segments 1 and 2 from midline, d with median layers of muscles removed. Anterior to left of figure (for muscle numbers, see (a, b) ***Weber*** 1928 and (c, d) ***Achtelig*** 1970).

*Chauliodes formosanus* has only one levator apodeme whereas *Corydalus* (and *Chauliodes* species examined by ***Kelsey***) all have two and (2) *C. formosanus* lacks the third of the group of three unguitractor muscles of *Corydalus*.

The majority of pterothoracic muscles of *Corydalus* are similar in the two segments (***Kelsey*** 1957) and ***Maki*** (1936) noted that muscle complements of abdominal segments 2–6 of *Chauliodes* also resemble each other closely, reflecting the undifferentiated skeletal structures of these segments. From segment 7 backwards, modifications of many dorsal, ventral, and ventral transverse muscles occur, and these are more pronounced in segments 8 and 9 of both sexes.

## Circulatory system

Knowledge of the circulatory system in Megaloptera relies entirely on a comprehensive study of *Sialis lutaria* by ***Selman*** (1965), and his earlier appraisal of blood cells of the same species (***Selman*** 1962). The following details are from his papers.

The adult heart, which has many similarities to that of the larva, has seven chambers with a pair of ostia in each of abdominal segments 1–8, and its terminal region in segment 8. Alary muscles support it laterally and the extreme lateral walls are attached to the dorsal longitudinal muscles by connective tissue. It comprises two regions: in segments 1–4 the adult heart is larger than that of the larva and in segments 5–8, it is smaller. In all, the valve openings between successive heart chambers are longer than in the larvae. The heart lies between enlarged dorsal longitudinal muscles in segments 1–4. The inner longitudinal muscles in the female protect the heart from pressure of the ovaries, which otherwise might hamper circulation. ***Selman*** suggested that the posterior part of the heart is probably functional in gravid females only when these muscles contract and reduce the upward pressure of the ovaries. Dorsal musculature, and the heart itself, are reduced in segments 5–8; the dorsal diaphragm is also greatly reduced in segment 8. The aorta, in the adult thorax, has no ostia or valves and is narrowed intersegmentally. It is especially narrow in the prothorax and narrows further in the head before broadening immediately behind the brain and dorsal to the pharynx. In this region there is also a pair of non-valvular lateral openings from the aorta immediately anterior and dorsal to the corpora cardiaca. A series of fine blood channels extend across the surface of the brain and have about the same height (10–15 μm) as the diameter of blood cells.

Circulatory organs occur in association with some of the head appendages. The antennae have pulsatile organs, and the maxillary and labial palpi have valvular structures which ensure unidirectional blood flow when the appendage is moved. The thoracic legs also have valvular circulatory organs (unlike some other insect orders, in which these are pulsatile). Wings are

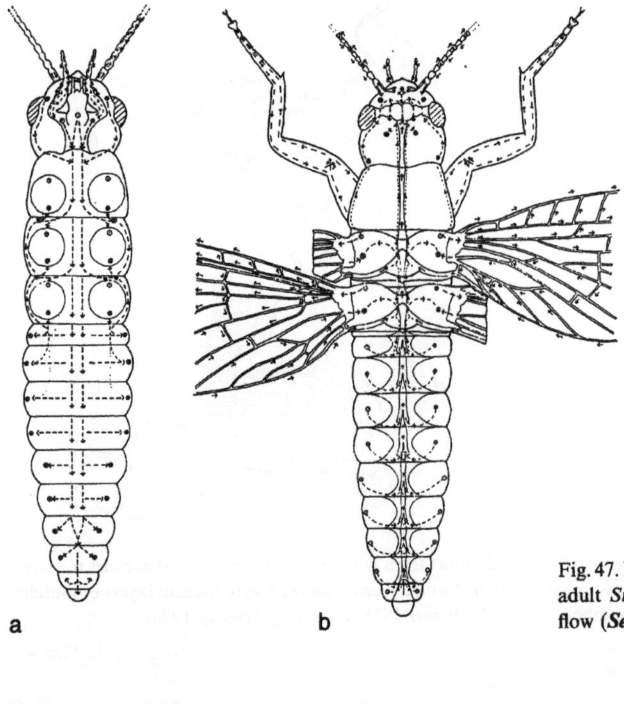


Fig. 47. Dorsal (b) and ventral (a) circulatory pathways in adult *Sialis lutaria*: arrows indicate direction of blood flow (***Selman*** 1965).

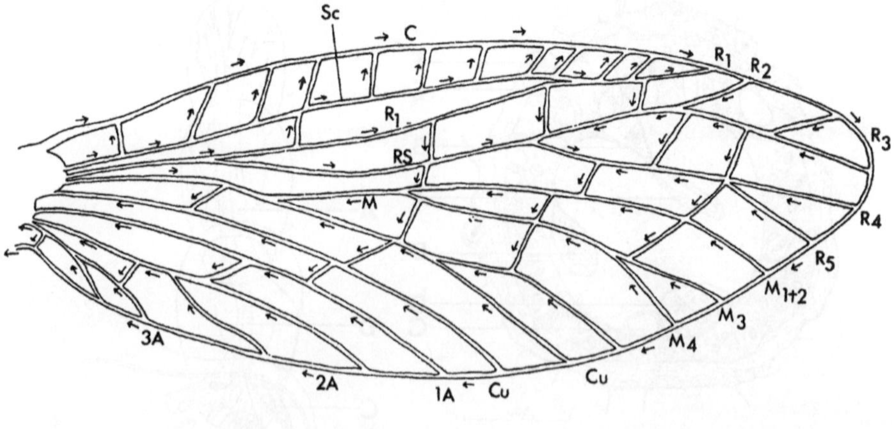


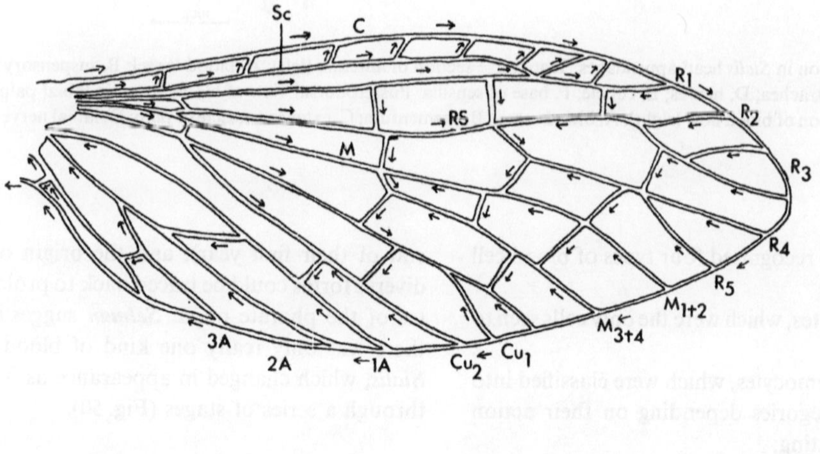


**Fig. 48. Blood circulation in wings of *Sialis*: direction of blood flow shown by arrows (*Selman* 1965).**

associated with two kinds of circulatory organs: pulsatile organs (scutellar organs, similar to those of many other insects but no blood vessel leads from these organs to the aorta as documented for some orders) and simple valvular structures, as pairs of ampullae for each wing, opening into the posterior part of the corresponding segment.

There is no dorsal or ventral diaphragm in the thorax of adult *Sialis*, and blood pathways there may be affected differentially in the two sexes by the gut diverticulum (or 'food reservoir'). In females this is restricted to the thorax, but extends into the first four abdominal segments in males. The abdomen is divided horizontally by both dorsal and ventral diaphragms, the latter being reduced in the more posterior segments. Some circulation pathways are shown in Figs 47–49. The main blood flow away from the head is ventral to the posterior tentorial arm, where several streams join. Most of this blood enters the prothorax ventrally and flow is restricted by the food reservoir. ***Selman*** (1965) summarised the flow pathway as: prothoracic leg – mesothoracic wing – mesothoracic leg – metathoracic wing – metathoracic leg – first abdominal segment. Blood flow in the wings (Fig. 48) follows the common insect pattern (see ***Arnold*** 1964) of outward along the more anterior veins and inward in veins towards the wing posterior. Blood from the anal veins discharges into the posterior ampulla, and thence to the thoracic haemocoel. Blood from the posterior wing margin passes into the scutellar organ through a one-way valve, and when the wing is extended virtually all blood passes through these organs which maintain a steady flow and the flow through the ampullae increased if some body movement in relation to the wings was allowed. No ampullar flow occurred with the wings folded.

Blood enters the abdomen near the ventral nerve cord and is driven backward by rhythmic contractions of the ventral diaphragm. Much of it flows laterally and dorsally to the heart in the first four abdominal segments, and there is very little blood flow in segments 9 and 10. The adult heart beat rate is about 51/minute at rest at 19 °C.

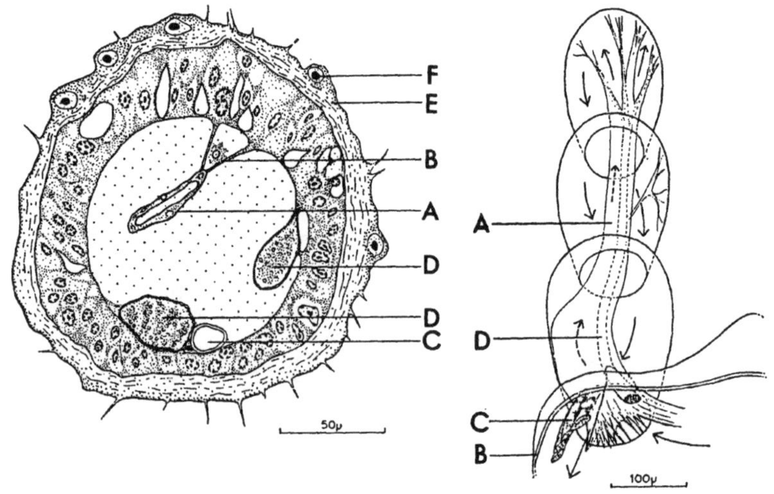


Fig. 49. Circulation in *Sialis* head appendages. Transverse section of antenna (left): A, blood vessel; B, suspensory fibres of blood vessel; C, trachea; D, nerves; E, cuticle; F, base of sensilla. Posterodorsal view of blood vessel in labial palp, arrows indicating direction of blood flow (right): A, blood vessel; B, prementum; C, extensor muscle of palp; D, labial nerve (***Selman*** 1965).

***Selman*** (1962) recognised four types of blood cell in *Sialis*:

1) proleucocytes, which were the only cells seen to divide;
2) hyaline haemocytes, which were classified into several categories depending on their action during clotting;
3) dark hyaline haemocytes, which were found only occasionally and take no part in clotting; and
4) granular haemocytes, which also take no part in clotting and frequently became enmeshed by the fine cytoplasmic threads of thigmotactic hyaline haemocytes.

Blood cells present are especially diverse in the adult at about a week old (and also in larvae at the end of their first year), and the origin of these diverse forms could be traced back to proleucocytes of the pharate pupa. ***Selman*** suggested that there was only really one kind of blood cell in *Sialis*, which changed in appearance as it passed through a series of stages (Fig. 50).

## Alimentary system

Early accounts of the alimentary canal of *Sialis* (***Dufour*** 1841, ***Loew*** 1848, ***Stitz*** 1926) have scarcely been augmented and, for example, ***Dufour***'s diagram was reproduced by ***Berland & Grassé*** (1951). The gut (Fig. 51) is straight, with the narrow oesophagus preceding a broader midgut

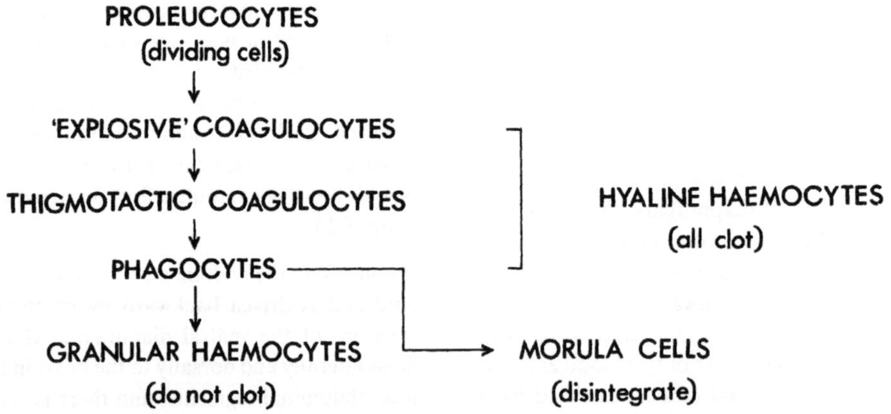


Fig. 50. Developmental history of blood cells in *Sialis lutaria* (***Selman*** 1962).

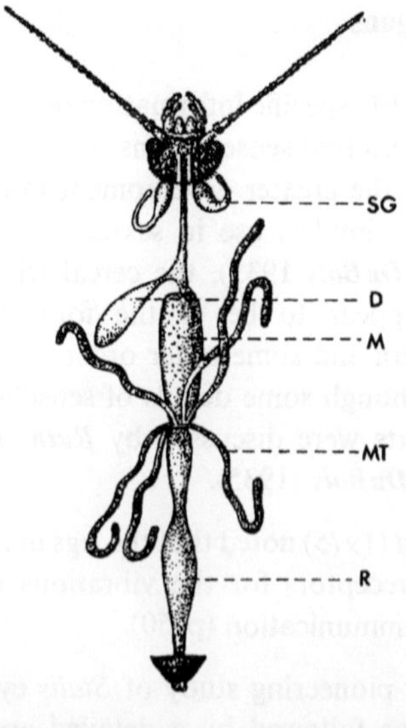


Fig. 51. Alimentary system of *Sialis lutaria*: D, foregut diverticulum or 'pouch'; M, midgut; MT, Malpighian tubules; R, rectum; SG, salivary glands (***Dufour's*** 1841 diagram as reproduced by ***Berland & Grassé***, 1951).

and having a diverticulum (or 'food reservoir') which is much longer in the male than in the female. There is a single pair of tubular labial salivary glands, and ***Röber*** (1941) also depicts mandibular glands in *Sialis*. The midgut has a deeply folded lining and a peritrophic membrane and is generally broader than the hindgut. In *S. rotunda* and *S. californica*, ***Azam & Anderson*** (1969) reported that the alimentary canal is not developed in the adult. In *Corydalus* the proventriculus has six small longitudinal folds covered with fine acanthae (***Boudreaux*** 1980).

There are six (*Sialis*: ***Suchow***'s, 1828, illustration showing only four is presumed erroneous) or eight (*Corydalus*: ***Leidy*** 1848) Malpighian tubules which are free and particularly long in *Sialis*.

## Respiratory system

Adult Megaloptera have two pairs of thoracic and eight pairs of abdominal spiracles, so that the respiratory system is typically holopneustic, in contrast to the larvae (p. 73).

The spiracular closing mechanism, described by ***Hassan*** (1944, *Sialis*), is complex, as the thoracic spiracles possess two different mechanisms. The anterior thoracic spiracle comprises an opening bounded posteriorly by a sclerotised bridge, the ventral edge of which is produced into a process (manubrium) to which an opening and a closing muscle are attached. Opposite the bridge, anterior to the opening, a large lever articulates with the upper end of the bridge, with the closing muscle running between the lever and the manubrium. An elastic membrane stretches between the manubrium and the two heads of the lever. When the occlusor (closing muscle) contracts, the lever moves downward and the manubrium upward so that the membrane covers the spiracular opening.

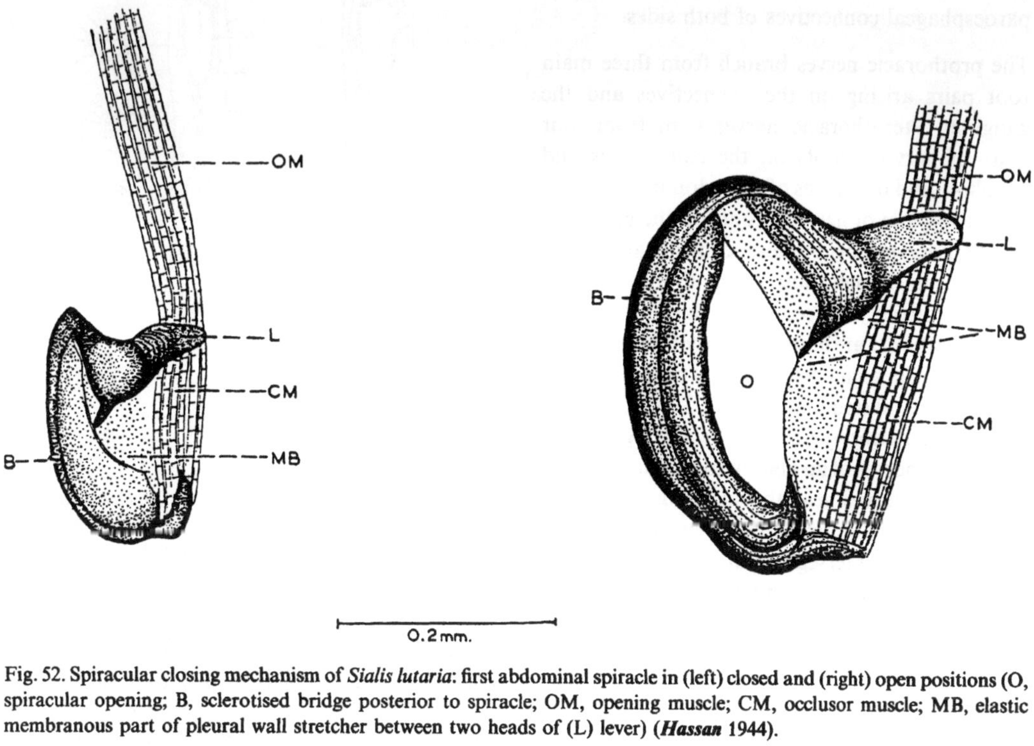


Fig. 52. Spiracular closing mechanism of *Sialis lutaria*: first abdominal spiracle in (left) closed and (right) open positions (O, spiracular opening; B, sclerotised bridge posterior to spiracle; OM, opening muscle; CM, occlusor muscle; MB, elastic membranous part of pleural wall stretcher between two heads of (L) lever) (***Hassan*** 1944).

This is shown for an abdominal spiracle in Fig. 52, and closely resembles the mechanism described in Mecoptera (***Hassan*** 1944).

## Nervous system

Several detailed studies have been made on the nervous system, but most of these deal with larvae. Although the larval and adult systems appear not to differ greatly, they are dealt with separately here. The general form of the brain of *Sialis* was claimed by ***Ehnbom*** (1948) to resemble that of Trichoptera and Lepidoptera. In *Chauliodes formosanus* adults, from the most detailed study available on any megalopteran (***Maki*** 1936), the central nervous system comprises the brain (in the medial region of the head and with the protocerebrum the largest region), the suboesophageal ganglion, three thoracic ganglia and eight abdominal ganglia. The eighth abdominal ganglion is larger than the anterior abdominal ones and close to the seventh; indeed, ***Loew*** (1848) apparently distinguished only seven ganglia in *Sialis*. The optic lobes of the protocerebrum are enlarged. The deutocerebrum is represented mainly as a pair of swellings bearing the antennal nerves and the tritocerebrum gives off (1) a pair of frontal ganglion connectives and (2) a pair of labial nerves (anterior) and paraoesophageal connectives (ventral) uniting the brain with the suboesophageal ganglion. The postoesophageal commissure of the tritocerebrum connects with the paroesphageal connectives of both sides.

The prothoracic nerves branch from three main root pairs arising on the connectives and the ganglion. Pterothoracic nerves stem from four pairs of nervous roots on the connectives and ganglion, and the nerves of an abdominal segment from two pairs of nervous roots on the ganglion. The first nerve roots of these segments are homologous even though those of the thorax arise on the connectives. The second and third nerve roots of the prothorax are homologous with those of the pterothoracic segments. Median nerves on the posterior of abdominal ganglia I–VII connect with the anterior region of the next ganglion. Transverse nerves of an abdominal segment are on the median nerves near the succeeding ganglion and connect with branches of nerves of that ganglion in the intersegmental region. They have branches to both dorsal and ventral nerves and to the nerves of adjacent ganglia. Transverse nerves in other segments may be otherwise modified.

## Sense organs

There is little specific information on many of the more generalised sense organs, such as antennae (in which the greater development in one sex of some taxa implies use in sexual attraction; see ***Geigy & DuBois*** 1935), the cercal trichobothria (which appear to be of the form typical of Neuroptera and some other orders) and mouthparts, although some details of sensilla on *Sialis* mouthparts were discussed by ***Rath*** (1888) and ***Geigy & DuBois*** (1935).

***Rupprecht*** (1975) noted that the legs of *Sialis* have mechanoreceptors for the vibrations utilised in sexual communication (p. 50).

*Eye*. The pioneering study of *Sialis* eyes by ***Ast*** (1920) was followed by a detailed appraisal of *Archichauliodes* (***Walcott & Horridge*** 1971), who pointed out that ***Ast*** had in fact described a novel kind of compound eye (in which there was movement of retinular cells during light-dark adaptation, a condition now known also in some other groups of insects) which he was, unfortunately, unable to appraise fully as his work was published posthumously with little discussion included.

Eyes in Megaloptera are large and hemispherical, and occupy a large proportion of the head area. In

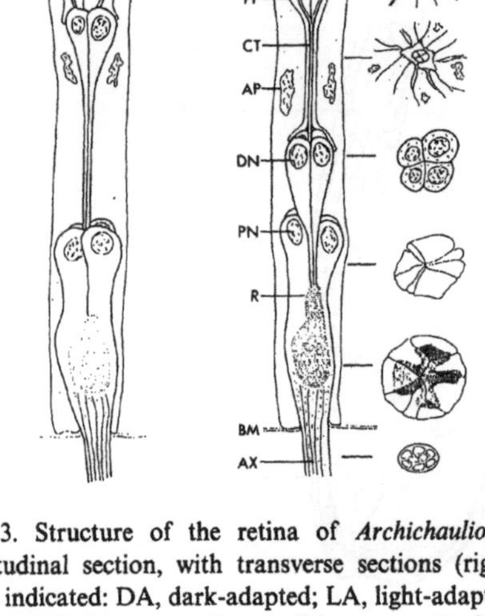


Fig. 53. Structure of the retina of *Archichauliodes* in longitudinal section, with transverse sections (right) at levels indicated: DA, dark-adapted; LA, light-adapted (C, cornea; CC, four tapered crystalline cone cells forming (CT) crystalline tract; DN, distal retinula cells; PN, four deeper, proximal retinula cells with nuclei; R, rhabdom, formed from parts of these eight cells; PP, primary pigment cells around each cone and crystalline tract; AP, accessory pigment cells between ommatidia; AX, eight axons; BM, basement membrane (***Walcott & Horridge***, 1971).

*Archichauliodes* the facets are rather flat and about 20 µm in diameter. Their general anatomical structure is shown in Fig. 53. Each corneal facet is subtended by a four-part crystalline cone with a distinct cone and outer zone. Mitochondria, microtubules and endoplasmic reticulum occur in the peripheral region. The cones taper over their length and are entirely surrounded by a pair of cells (primary pigment cells) which are less dense than the cone cells and have their nuclei about one third of the distance down the cone. The primary pigment cells contain many microtubules running parallel to the long axis of the cone cells, and many small dense particles. They are surrounded by 14–18 polygonal accessory pigment cells which fill the space between the cones, and each of these serves two ommatidia. The cone cells taper to a dense crystalline tract packed with longitudinal microtubules, and the length of the tract varies in length with the state of light/dark adaptation (Fig. 53) and ends in the region of four distal retinula cells which are, in turn, more deeply joined by four proximal cells. The fused rhabdom is constructed from all eight of these cells, and there are eight axons below the basement membrane. In other types of insect eye, extensions of the cone usually reach the basement membrane (***Walcott & Horridge*** 1971).

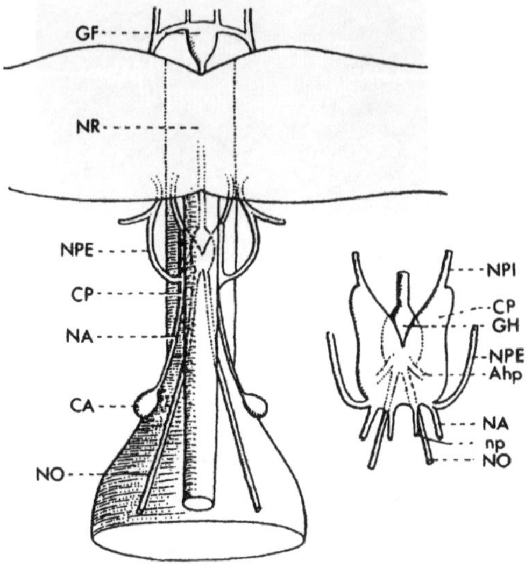


Fig. 54. Retrocerebral complex of *Sialis lutaria* adult; left, dorsal aspect of complex; right, detail of hypocerebral ganglion and corpora cardiaca: Ahp, hypocerebral/corpus cardiacum anastomosis; CA, corpora allata; CP, corpora cardiaca; GF, frontal ganglion; GH, hypocerebral ganglion; NA, allatal nerve; NO, oesophageal nerve; np, aortic nerves; NPE, NPI, external and internal paracardiac nerves; NR, recurrent nerve (***Cazal*** 1948).

### Neuroendocrine system

***Cazal*** (1948) depicted the general appearance of the retrocerebral endocrine complex of *Sialis lutaria* as in Fig. 54, and the major glands had earlier been noted by ***Hanstrom*** (1940). It is not unusual except for the position of the corpora allata which are closely associated with the sides of the oesophagus inside the prothorax. The hypocerebral ganglion is small and gives off two oesophageal nerves. The corpora cardiaca are fused posteriorly and are near the lateral anterior openings of the aorta (***Selman*** 1964).

No details of the neurosecretory cells of adult Megaloptera are available, but those of larval *Sialis* were appraised by ***Panov & Davydova*** (1976), and a detailed description of the system given by ***Rahm*** (1952).

## Cytology

***Klingstedt*** (1937) appraised the limited early information on chromosomes of Megaloptera in relation to other neuropteroids. Data on only two taxa, both Corydalidae, were available: *Chauliodes japonicus* (2n = 20 [i.e. 18 + X + Y]) (***Itoh*** 1933a) and *Protohermes grandis* (2n = 23 [i.e. 20 + ∝ + B1 + B2]) (***Itoh*** 1933b). No evidence of distance conjugation was found, although this is widespread in Planipennia (***New*** 1989) and Raphidioptera (***Aspöck & Aspöck*** 1971). The male is the heterogametic sex and sex determination appears to be XY/XX, as in many neuropteroids, in *C. japonicus*. However, in *Protohermes* (where female data are not available), the sex chromosomes may be $XY^1Y^2$ as the three smallest chromosomes seem to lack structurally homologous mates (***Hughes-Schrader*** 1980).

Data on the chromosomes of two other Corydalidae (*Corydalus cornutus*: 2n = 24, and *Neohermes filicornis*: 2n = 22) were provided by ***Hughes-Schrader*** (1980). The sex chromosomes (XY/XX) invariably formed a bivalent and segregated synchronously with the autosomes. The sex-bivalent is of the 'parachute' type in *Neohermes*. These megalopteran meiocytes were distinguishable from those of previously described Neuroptera and Raphidioptera also by absence of autosomal univalents and of the 'straight-jacket' deformation and chromosomes in individual spindle units, characters which may be phylogenetically significant. In *C. cornutus*, the sex chromosomes (Fig. 55) and the smallest pair of autosomes are minute, with Y being smaller than X in the male.

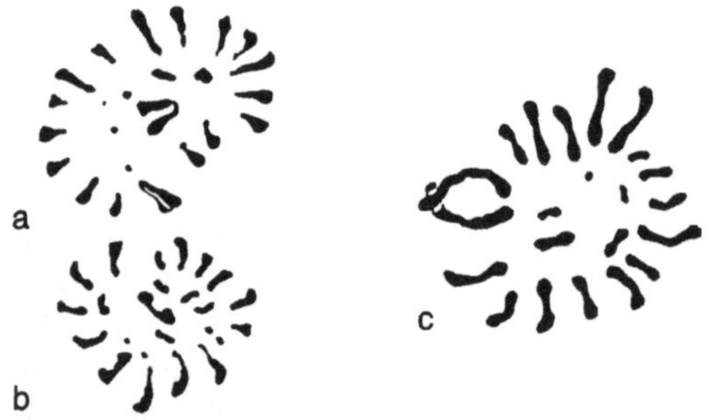

Fig. 55. Chromosomes of (a, b) *Corydalus cornutus* and (c) *Neohermes filicornis*; a, spermatogonial metaphase; X and Y are the minute left centre chromosomes, small autosomes below these; b, follicular cell of ovary, two Xs at lower right centre; c, spermatogonial metaphase, sex chromosomes upper right centre (***Hughes-Schrader*** 1980)

Other chromosomes are medio- or submedio-kinetic and the arms of most major autosomes are differentiated – which ***Hughes-Schrader*** noted as 'a curious feature of uncertain significance'. All autosomes of *N. filicornis* (Fig. 55) are medio-kinetic, none is really minute, and structural differentiation of the two arms is lacking. The sex chromosomes, again, are the smallest of the set, with Y minute. Male meiosis closely parallels that of *Corydalus*.

## Courtship and Mating

Megaloptera are bisexual and parthenogenesis has not been confirmed in the order. Sex ratios have only rarely been noted, but ***Pritchard and Leischner*** (1973) noted that male *Sialis cornuta* predominated by about 2:1 (emergence cages) or 1.5:1 (shoreline pitfall traps for fully grown larvae).

Mating, and its preliminary behaviour, has been described by several workers for *Sialis* (***Stitz*** 1928, ***Killington*** 1932, ***Symes*** 1935, ***Geigy & DuBois*** 1933, ***Seitz*** 1941, ***Azam & Anderson*** 1969, ***Pritchard & Leischner*** 1973). Accounts differ in the amount of detail provided but indicate that the process is rather similar in the species which have been studied. Males and unmated females of *S. lutaria* and *S. fuliginosa* communicate by reciprocal signalling (***Rupprecht*** 1975) and three kinds of abdominal signalling occurred. Females and males rhythmically vibrate their abdomens vertically (Fig. 56), and females of *S. cornuta* tend to remain still whilst doing this and males are more active and may vibrate whilst walking around (***Pritchard & Leischner*** 1973). These vibrations allow mutual approach and recognition of species and sex. Secondly, ***Rupprecht*** noted that males

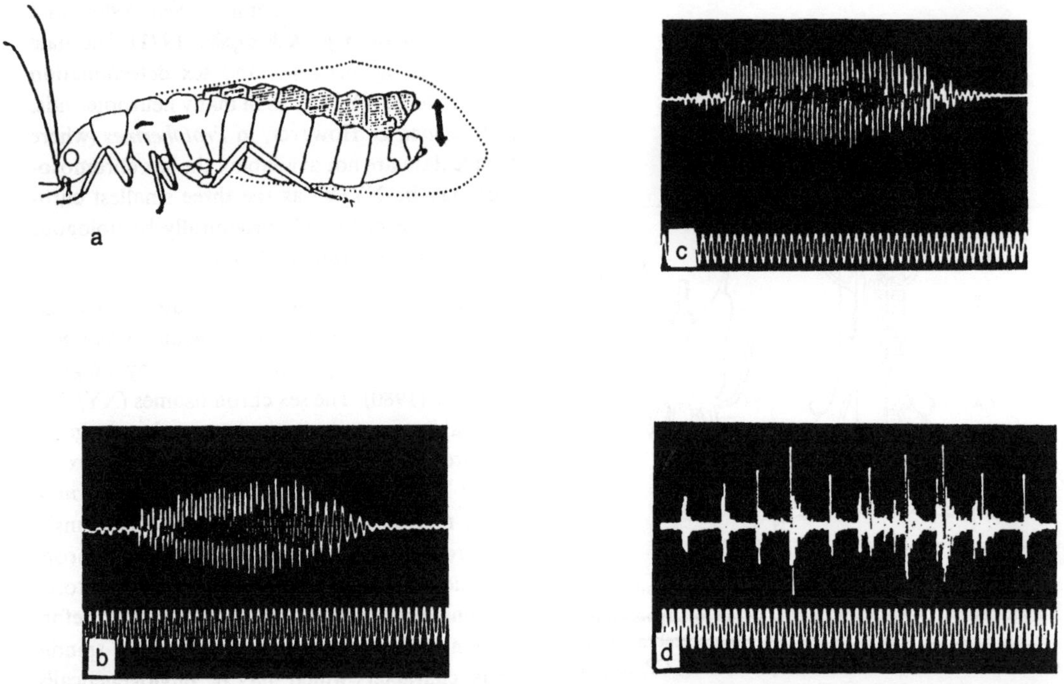

Fig. 56. a. Abdominal 'tapping' behaviour of *Sialis* during courtship; b, c, d, recordings of: b, rhythmical vibratory signal of a male *S. fuliginosa*; c, vibratory signal of a female *S. fuliginosa*; d, tapping signal of female in response to signals of a male *S. fuliginosa*. Scale frequency 100 Hz (***Rupprecht*** 1975).

may undertake prolonged abdominal vibration, and the third kind of signal involved tapping of abdomen and wings on the ground. The pattern of this differed substantially between species and may constitute a specific display. It occured only in males of *S. lutaria*, and was interpreted as 'feedback' to females after their approach, but was done by both sexes of *S. fuliginosa* during the first phases of their mutual approach. The vibrations are transmitted through the legs to the substrate and then to the legs of the other individual.

When the two sexes of *S. cornuta* meet they face each other and both individuals vibrate their antennae. After this reciprocal dueting the female turns and walks away from the male (***Pritchard & Leischner*** 1973) but, when she stops or slows down, the male starts to vibrate his antennae and mouthparts over the female's wingtips, hindlegs and underside of the abdomen. It is possible that the notched labrum of the male may play a sensory role in communication at this stage, and this behaviour is an immediate prelude to mating if the female is receptive.

The male pushes underneath the female's abdomen and pushes upwards, then folding his wings to one side and twisting his abdomen upwards to copulate (Fig. 57). A receptive female may be surrounded by several males, attracted by a pheromone, but she copulates only once (***Elliot*** 1977).

Mating in *Sialis* lasts only for several minutes, during which a large white spermatophore is transferred to the females. The spermatophore of *S. lutaria* (Fig. 58) is about 2 mm long and vase-shaped (***Khalifa*** 1949). It contains two large sperm sacs each ending in a long narrow blind tube. The slender end is inserted into the female's bursa at mating, and the whole structure hardens rapidly on exposure to air. The main body of the spermatophore is taken gradually into the genital opening where the muscular walls of the bursa aid escape of sperm. In other cases the female has been observed to chew on the spermatophore soon after mating has finished. She may then feed on it, but mandibular action may also stimulate or aid release of sperm. Remains of old spermatophores are also removed and eaten by the females.

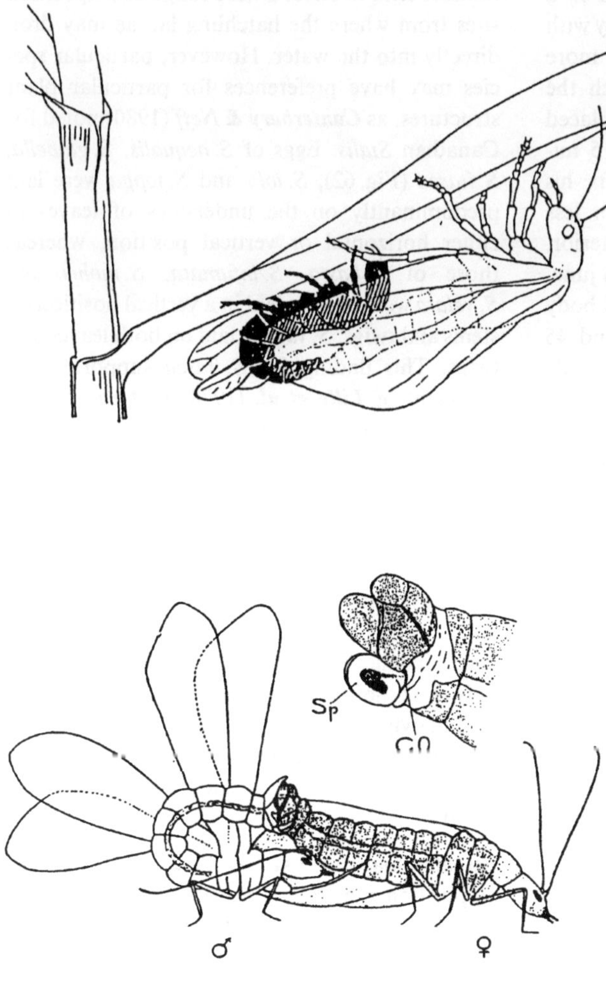


Fig. 57. Copulation in *Sialis*: (top) a mated pair, male (black) the smaller; (bottom) mating, with insert of female abdomen gripping spermatophore (Sp) in copulatory opening (CO); (top from ***Seitz*** 1941, bottom from ***DuBois & Geigy*** 1935).

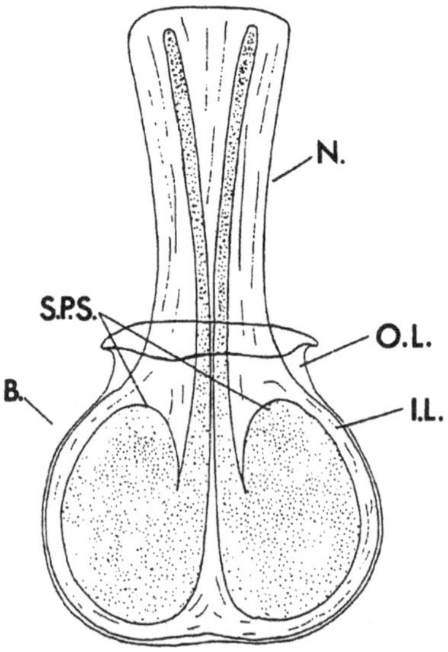


Fig. 58. Spermatophore of *Sialis lutaria*: B. body; N. neck; I.L., O.L., inner and outer layers; S.P.S., sperm sacs (***Khalifa*** 1949).

Mating of *Corydalus cornutus* in captivity was described by ***Parfin*** (1952) and she, also, noted that more than one male can be attracted to a virgin female and males then fight ferociously with their mandibles. Mandibles also play a more direct role in courtship and, together with the antennae, are used in initial contact and placed over the female's wings. After about 5–15 minutes (two cases observed) the male lifts his mandibles and, with his head still more or less parallel to that of the female but more posterior, bends his abdomen at the thorax-abdomen junction to about a right angle with his normal body axis, and copulates. Mating is brief (30 and 45 seconds in those observed), after which the male resumes the 'mandibles-on-wing' posture but will still physically repulse other males which approach. A second copulation was attempted by one male and these laboratory observations took place on a horizontal cage floor: it is not known if this reflects a normal mating position. No spermatophore was recorded, as for *Nigronia serricornis* (***Needham & Betten*** 1901) but it is not clear if it is really absent. In *N. serricornis*, one female observed by ***Needham & Betten*** may already have laid some of her eggs before being seen to mate.

The use of mandibles in this way differed from earlier suggestions (***Riley*** 1873, ***Walsh & Riley*** 1868) that male *C. cornutus* use these structures to grasp the female.

The overall role of scents in megalopteran courtship is not wholly clear, though seemingly well established in *Sialis* (***DuBois & Geigy*** 1935). ***Evans*** (1972) implied that mating behaviour could differ between Corydalinae and many Chauliodinae, with the male of the former producing a strong female-attractant scent from the eversible glands at the abdominal apex. Chauliodine females may play a more active role, although males of some (*Dysmicohermes, Orohermes*) have scent glands.

## Oviposition and fecundity

Eggs are laid soon after mating has finished, usually within a day, and female *Sialis* search for an oviposition site shortly after feeding on the spermatophore. They lay eggs on vegetation, bridges, or other objects overhanging water (***Azam & Anderson*** 1969) and *S. cornuta* commonly lays on the undersides of trunks and roots of uprooted trees (***Pritchard & Leischner*** 1973). In the laboratory this species laid on undersides of logs, away from the light. *S. lutaria* oviposits on plants and stones near water (***Syms*** 1935) and the habits of *Corydalus* and *Nigronia* are similar (***Baker & Neunzig*** 1968), so that members of both families tend to select a wide range of oviposition sites from where the hatchling larvae may drop directly into the water. However, particular species may have preferences for particular plant structures, as ***Canterbury & Neff*** (1980) noted for Canadian *Sialis*. Eggs of *S. aequalis, S. glabella, S. itasca* (Fig. 62), *S. iola* and *S. joppa* were laid predominantly on the undersides of leaves in either horizontal or vertical position, whereas those of *S. hasta, S. infumata, S. mohri* and *S. velata* were usually laid in a vertical position on stems and twigs. *S. vagans* laid on both leaves and twigs. This finding for *S. itasca* supported the comment of ***Lilly et al.*** (1978) that egg masses were found 'on the underside of hornbeam and buckeye leaves 0.5–3 m above the water surface'. Preparation for egglaying for *Sialis* includes secretion of an adhesive substance from the abdomen onto the potential oviposition substrate.

Those Megaloptera which have been documented appear to lay all their eggs together in one batch, and a female *Sialis* lays her full complement in only 1–2 hours (***Canterbury & Neff*** 1980). A second oviposition can occur in *S. lutaria* (***DuBois & Geigy*** 1935) but this may more usually reflect disturbance of a laying female (***Canterbury & Neff*** 1980).

Females may lay batches close together if an oviposition substrate is restricted. In an extreme case, ***Azam & Anderson*** (1969) noted that certain branches projecting over water were 'encrusted' with batches of *S. californica* eggs for about a metre.

Counts of eggs in the ovarioles of newly-emerged females of most of a range of *Sialis* species are very similar to those of eggs in a mass, strongly supporting opinion that a single oviposition is usual. As examples, *S. cornuta* females contained 615 ± 29 eggs and laboratory-laid batches averaged 535 ± 46 eggs (***Pritchard & Leischner*** 1973), and data accumulated by ***Canterbury & Neff*** (1980) are summarised in Table 1. One anomaly here is *S. aequalis*, for which ***Woodrum & Tarter*** (1973) found females to contain 657 (402–818, n = 10) eggs, but batches (only two examined), 262 and 335 eggs. This species appears to be exceptional. Fecundity in *Sialis*, therefore, is generally equivalent to the number of eggs in females, soon after they emerge, and fecundity varies severalfold between species. *S. joppa*, for example averaged only 187 eggs/batch in contrast to *S. mohri* with 718 eggs/batch (Table 1).

Dissection of a female *C. cornutus* approximately three days old revealed eggs which appeared to be close to complete development (***Parfin*** 1952), and a similar pattern may, thus, hold for members of this family. In general in Sialidae, fecundity of lake and pond species is greater than that of stream species (***Canterbury*** 1978). Egg batches can be larger in Corydalidae than Sialidae, and ***Riley*** (1877) recorded batches of *C. cornutus* as having 2000–3000 eggs each. ***Baker & Neunzig*** (1968) gave an average figure of 1080 eggs for this species, and ***Brigham*** (1982) noted 'about 1300 eggs', 760 eggs for *N. serricornis*, 630 for *N. fasciatus*, and 525 for *Chauliodes pectinicornis*. The latter usually produces two separate egg masses during a period of oviposition, and ***Brigham*** (1982) claimed that *C. cornutus* usually lay two egg masses and occasionally produce a third batch.

## Eggs and batch formation

Egg batches or masses are formed as successive eggs are deposited by a female, and batches may differ in appearance between species, depending on the size and degree of motility/agility of laying females. Thus, *Sialis cornuta* lays eggs in rows, one row at a time and moves backward over the growing mass as each row is completed (***Pritchard & Leischner*** 1973). *S. rotunda* and *S. californica* deposit a row of eggs by swinging the abdomen left and right of the midline and, after depositing a row, move forward before commencing the next one. The number of eggs in a row is affected by the regularity and convexity of the substrate, and ranges from 4 or 5 to 20–25. Egg masses of *S. rotunda* are narrower than those of *S. californica*, as the female is smaller and cannot swing her abdomen as far to the sides, but also because she bends the abdomen vertically to lay 'upright' eggs (***Azam & Anderson*** 1969). In *S. lutaria*, where rows typically contain 10–20 eggs (***Davis*** 1903) and the female moves backward whilst laying, the first row acts as an abutment for the second, and so on. Later rows are not laid directly from one side to the other but, rather, from one side to the midline and thereafter from the other side to the midline (Fig. 59). Eggs were laid at rates of 8–21/minute. As each egg is laid, a drop of cement precedes it.

As ***Canterbury & Neff*** (1980) emphasised, *Sialis* have two rather different patterns of egg deposition, the pattern sometimes being characteristic of a given species. Eggs are placed either (1) in a horizontal position with the longitudinal axis of the egg parallel to the sustrate, or (2) in a vertical or upright position, with the longitudinal axis perpendicular to the substrate or inclined to it at an acute angle.

*Sialis* egg-batches are unilaminar and, although there may be cement on and between the eggs, not

Table 1. Numbers of eggs in egg masses of North American *Sialis* species, compared with numbers of eggs in ovaries of newly-emerged adults. From Canterbury & Neff, 1980, including data on *S. aequalis* from Woodrum & Tarter, 1973, and on *S. itasca* from Lilley et al., 1978.

| Species | No. of eggs in egg masses | | | No. of eggs in ovaries | | |
|---|---|---|---|---|---|---|
| | N | Average | Range | N | Average | Range |
| *S. hasta* | 6 | 334 | 227–477 | 3 | 384 | 305–452 |
| *S. iola* | 6 | 479 | 289–655 | 2 | 353 | 283–423 |
| *S. joppa* | 4 | 187 | 170–203 | 3 | 186 | 158–230 |
| *S. infumata* | 4 | 444 | 440–451 | 3 | 480 | 445–550 |
| *S. aequalis* | 2 | 298 | 262–335 | 10 | 657 | 402–818 |
| *S. vagans* | 3 | 479 | 437–573 | 3 | 437 | 329–498 |
| *S. glabella* | 3 | 242 | 192–312 | 1 | 225 | – |
| *S. mohri* | 3 | 718 | 500–905 | 5 | 574 | 463–785 |
| *S. velata* | 2 | 679 | 577–780 | 4 | 438 | 340–533 |
| *S. itasca* | 3 | 560 | 330–583 | 3 | 539 | 454–587 |

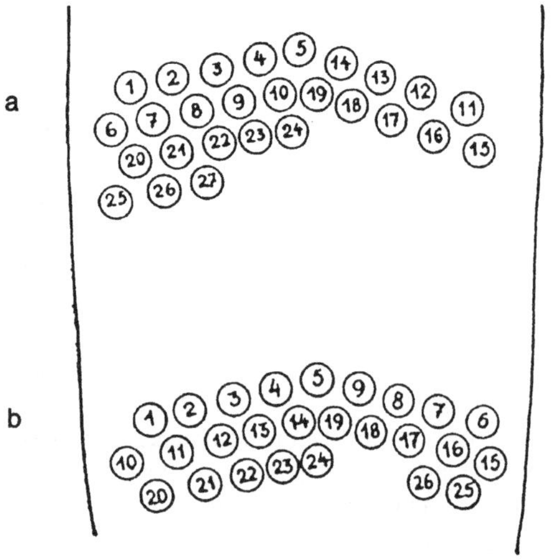


Fig. 59. Egg batch formation of *Sialis*: consecutive numbers are sequence of egg depositions of two batches (***Seitz*** 1941).

otherwise covered. Some Corydalidae egg batches, in contrast, comprise several layers of eggs with a distinct covering. Up to five layers occur in *Protohermes grandis* (***Miyakawa*** 1979). Whereas ***Riek*** (1970) commented that egg batches of sialids on rocks resembled lichens in appearance, ***Riley*** (1877) likened the appearance of *C. cornutus* batches (which average about 21 mm diameter) to 'the droppings of a large bird', as the convex batch is covered with an irregular chalky white covering which may be present also irregularly on the substrate (usually a leaf) around the batch. The batch consists of three layers of eggs (Fig. 60). The two lower layers are horizontal and the eggs form

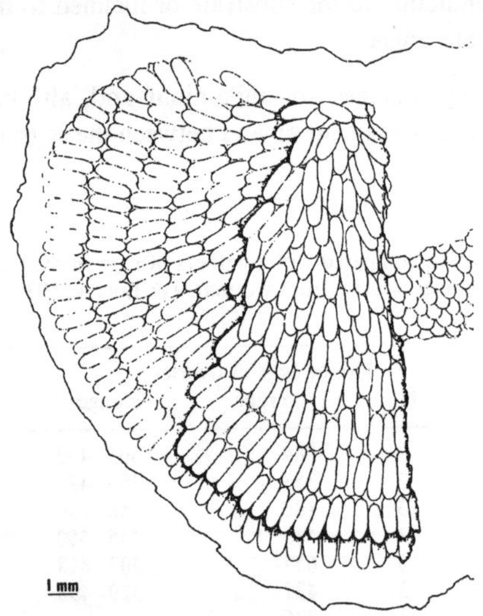


Fig. 60. Egg batch of *C. cornutus*: overlying tough white cover partially removed, and bottom, middle and top layers of eggs shown from left to right (***Baker & Neunzig*** 1968).

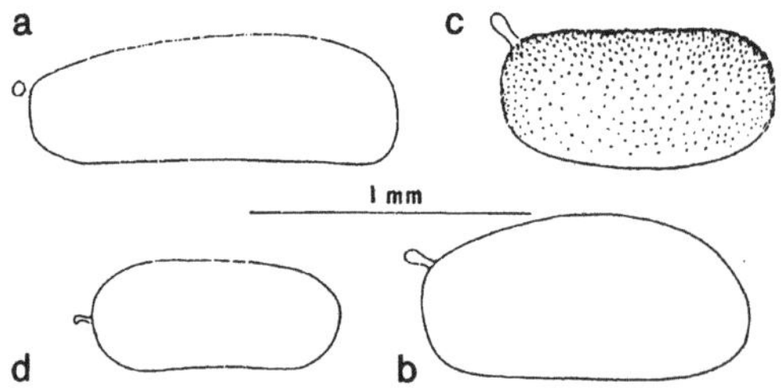


Fig. 61. Eggs: lateral views of eggs of a, *Corydalus cornutus*; b, *Neohermes concolor*; c, *Chauliodes pectinicornis*; d, *Nigronia serricornis* (***Baker & Neunzig*** 1968).

semicircular arcs. Eggs in the top layer are more irregular. They are horizontal around the periphery of the batch but more irregular and approaching a vertical position near the batch centre and anterior (***Baker & Neunzig*** 1968) (Fig. 60). Batches of *N. serricornis* and *N. fasciatus* usually have two layers of eggs with the upper one being much the smaller (20–50 eggs). The eggs have surface droplets of a viscous liquid, which persists throughout egg development. Batches of *Chauliodes pectinicornis* have a single layer of eggs without any covering. *Neohermes concolor* also have one egg layer, but the batch has a soft thin brown covering (details from ***Baker & Neunzig*** 1968).

## Egg morphology

Within a batch eggs are rather uniform in size, but there are substantial size differences between species. *Sialis joppa* has the smallest eggs (0.50 mm length, 0.19 mm diameter) and *S. glabella*, the largest (0.65 mm length, 0.22 mm diameter) of the species examined by ***Canterbury & Neff*** (1980), and there are slight size variations between localities for the same species. Egg colour also varies interspecifically, ranging from white to dark brown, and ***Canterbury & Neff*** (1980) observed that horizontal eggs tend to be pale (white, yellow, tan) and those in vertical positions dark (grey, reddish orange, brown). Eggs darken during development.

Corydalid eggs are larger than sialid eggs. *C. cornutus* eggs are 1.33–1.43 mm in length and 0.39–0.58 mm in diameter, for example. The largest detailed are *Nigronia serricornis*, with egg length up to 1.67 mm and diameter up to 0.59 mm (***Baker & Neunzig*** 1968), but *Neohermes concolor* eggs are only about 0.85 × 0.40 mm. The shape of the eggs differs in different taxa of Corydalidae (Fig. 61), apparently to a greater extent than in Sialidae.

Fig. 62. Eggs: chorionic structure and oviruptor of *Sialis* spp.: A, *S. joppa*, with micropylar projection (A); B, *S. itasca*, with micropylar projection (A) and oviruptor (B) (***Canterbury & Neff*** 1980).

A number of early descriptions of megalopteran eggs (for examples, by ***Needham & Betten*** 1901, ***Davis*** 1903, ***Geigy & DuBois*** 1935, ***Hamilton*** 1940) are superficial, although providing useful information. Thus, ***Davis*** (1903) demonstrated that a protruding micropylar 'turret' occurs on the eggs of representatives of all three main groups in the order. This prominent structure is cylindrical or fusiform in *Sialis* (***Canterbury & Neff*** 1980), and the apex can be enlarged and globular in Corydalidae (Fig. 61) where the micropyle can be displaced laterally from the true apparent anterior pole of the egg so that it protrudes above the batch (or layer) surface. *Leptosialis* seemingly resembles Corydalidae, from the figure given by ***Crass*** (1949).

The chorion of *Corydalus* eggs is smooth, as it is in some non-covered batch taxa such as *Nigronia* and *Protohermes* (***Miyakawa*** 1979). In contrast, eggs of *Chauliodes pectinicornis* have numerous small peltate tubercles on the upper surface; these are reduced laterally and absent on the lower surface (***Baker & Neunzig*** 1968).

*Sialis* eggs are also partially or completely ornamented. The ornamentation of both chorion and micropyle, and the shape of the latter, differ interspecifically (***Canterbury & Neff*** 1980, Fig. 62). Ornamentation consists of raised peltate structures. They may function as 'anchors' for secretions from the female's accessory gland, and there is some suggestion that they could aid in establishing a plastron for respiration when eggs are wetted or submerged (***Canterbury & Neff*** 1980), although ***Hinton*** (1981) did not know whether air was actually held by the tubercles. The projections are larger and more prominent on vertical eggs, on which they are also sparser and smaller near the egg base. ***Canterbury & Neff*** pointed out that this pattern could be consistent with a 'plastron' function, because vertical eggs tend to be laid on substrates exposed to wetting, and horizontal eggs are commonly in more protected areas where, although subject to condensation, they might avoid direct precipitation. Micropylar projections also vary, and may serve to define the outer boundary of a surface water film and facilitate aerial respiration. Vertical eggs tend to have long cylindrical or pedunculate projections, and the spiny apex may also facilitate carrying sperm in suspension to the subapical micropylar openings (***Canterbury & Neff*** 1980).

## Embryonic development

Until recently, understanding of embryonic development in Megaloptera has been poor and derived solely on data from one species, *Sialis lutaria*, for which ***Strindberg*** (1915) provided a histological study of germ layer formation and organogenesis, and ***DuBois*** (1936, 1938) undertook an experimental study indicating the main embryonic changes during development. This scanty framework has now been augmented considerably by studies on Japanese species *Sialis mitsuhashii* (***Suzuki et al.*** 1981, 1987, ***Ando et al.*** 1985) and

*Protohermes grandis* (***Miyakawa*** 1979, 1980), so that representatives of both families have now been appraised in some detail. Nevertheless, as ***Ando et al.*** (1985) commented, the amount of embryological information available is poor in comparison to many other insect orders.

The earliest development of *S. mitsuhashii* is summarised in Table 2 (***Suzuki et al.*** 1981, from which the following details are taken). When the egg is laid the chorion is soft and underlain by a thin vitelline membrane. In the central dorsal region (Fig. 63) the thin periplasm thickens to form an 'island' of cytoplasm where the nucleus undergoes the first maturation meiosis. Two types of yolk globules occur: (1) Type-I are 2–17 μm in diameter and stain with eosin, and the smaller ones occur around the periphery. Larger globules (> ca 12 μm) contain small granules; (2) Type-II are ca. 7–17 μm in diameter and contain granules which stain with haematoxylin. Type-I globules are predominantly on the dorsal side, and Type-II on the ventral side, but they may sometimes rather be more anterior (Type-I) or more posterior (Type-II). Some eggs in a batch have almost wholly Type-I globules, and this seems to be the only type of yolk in *S. lutaria* (***DuBois*** 1938).

Table 2. Early embryonic development of *Sialis mitsuhashii*. From Suzuki et al., 1981.

| Time (hr) | State of Development |
|---|---|
| 0–0.25 | 1st meiosis. Male pronucleus in yolk. |
| 0.33–0.50 | 2nd meiosis. Polar bodies formed. |
| 1 | Fertilisation and 1st cleavage |
| 2 | 2nd cleavage |
| 4 | 3rd cleavage |
| 6 | 4th cleavage. Some energids to posterior half of egg, distributed irregularly. |
| 9 | 5th cleavage. |
| 12 | 7th cleavage. Centrifugal migration of energids begins. |
| 13 | 8th cleavage. Mitotic rate of nuclei differs in different parts of egg. |
| 14 | 9th cleavage. Energids almost reach egg periphery. Polar granules/polar bodies disappear at 12–14 hr. Syncytial blastoderm begins. |
| 15–16 | 1st division of syncytial blastoderm nuclei. About a dozen yolk cells remain in yolk. |
| 20 | Cellular blastoderm forming. Yolk cells increase in number. |
| 22 | Blastoderm cell membrane formed, cellular blastoderm completed. 2nd division of blastoderm cells. Differentiation of embryonic and extraembryonic areas. |
| 26 | Median plate differentiates as spot in centre of ventral plate. |
| 29 | Median plate elongates, occupies two-thirds of ventral plate. |
| 32 | Primitive groove appears. Invaginating median plate becomes multicell-layered, especially caudally. |

Numerous large vacuoles occur, and sperm are often present in the central micropylar canal. The second maturation division occurs in the same cytoplasmic island as the first, and finishes in 20–30 minutes after oviposition. The female pronucleus and the three polar bodies are produced, and the former moves from the anterior pole to about one third of the egg length where the male pronucleus is located. Fertilisation occurs near there one hour after oviposition. Cleavage occurs rapidly, the second cleavage follows about two hours after egg deposition, and the third and fourth cleavages at successive intervals of about two hours. Further development follows the timetable indicated in Table 2. At about 26 hours several cells on the midline near the centre of ventral plate (Figs 63, 64) become taller than their neighbours, marking the initial formation of the median plate, which begins to extend anteriorly and posteriorly at about 27 hours as the plate becomes more evident.

The primordial germ cells are foreshadowed by a mass of about six cells near the caudal end of the ventral plate at about 25 hours, and the ventral plate occupies some two-thirds of the egg surface by 31 hours, when the protocephalon differentiates in most of the anterior region to form the germ band (Fig. 64). Cell division now occurs more frequently in the lateral plates than in the median plate and, as development proceeds, the germ band decreases in size on the yolk (Fig. 65). The primitive grooves occur as shallow invaginations at about 32 hours. The early formation on the median plate occurs also in some Neuroptera.

Observations on development of later embryos of *S. mitsuhashii* were discussed by ***Ando et al.*** (1985) and these are shown in Figs 64, 65. Metamerisation is apparent at two days, with four gnathal, three thoracic and two abdominal segments, and a large caudal lobe flexed dorsally across the posterior of the egg. By day three, the embryo has elongated as the abdomen grows larger so that its tip extends further towards the anterior pole of the egg. Metamerisation is then distinct except at the apex of the abdomen. The protocephalon shows the primordia of the labrum, antennae and stomodaeal pit, and the mandibles, maxillae and labia are also recognisable as paired swellings. The thorax segments bear small medioposterior leg buds, but no abdominal swellings are visible. At day four the embryo reaches its maximum length, and the appendage rudiments continue to develop. The labial primordium becomes bilobed, antennal and mandibular ones are more prominent and the maxillary lobes have two developmental points. Thoracic leg rudiments are

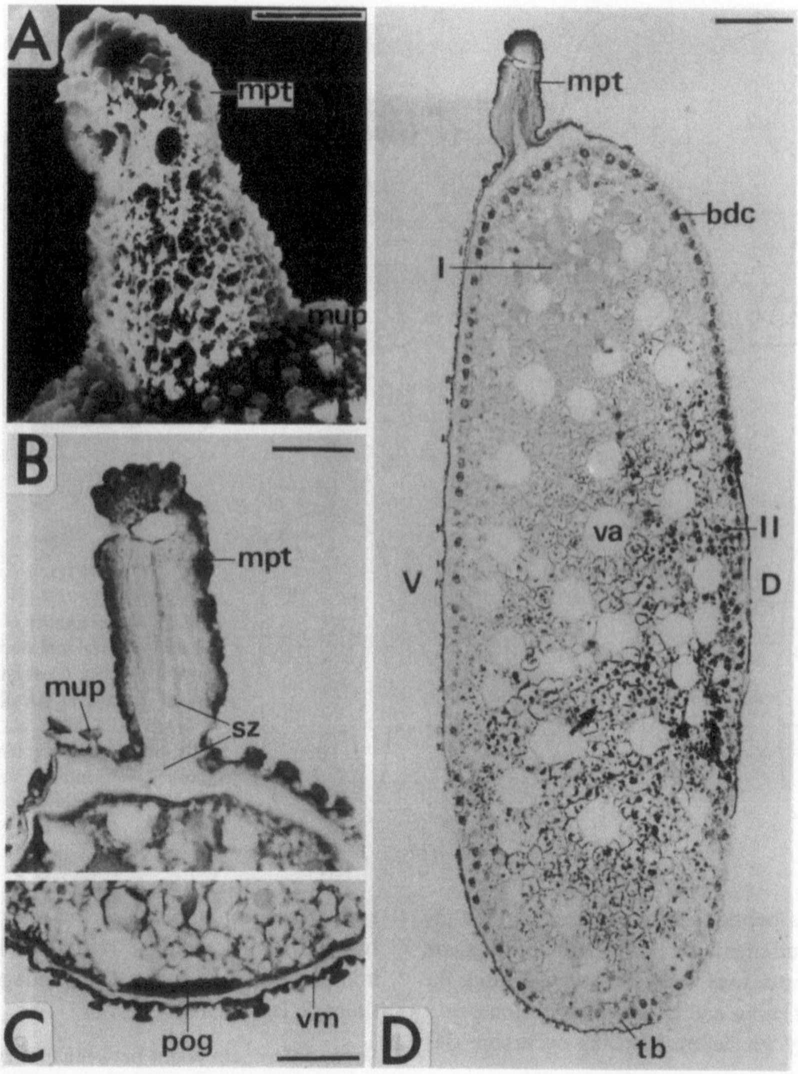


Fig. 63. Embryology of *Sialis mitsuhashii*: A, micropylar projection; B, same, longitudinal section just after oviposition; C, 20-minute-old egg, longitudinal section showing polar granule; D, 25-hour-old egg, sagittal section: D, dorsal; V, ventral; bdc, blastoderm cell; mpt, micropylar projection; mup, mushroom-shaped projection; pog, polar granule; sz, spermatozoa; tb, thickening of blastoderm; va, vacuole; vm, vitelline membrane; I, II, type-I and type-II globules; scales: 20 µm (A–C), 50µm (D) (from ***Suzuki*** et al. 1981).

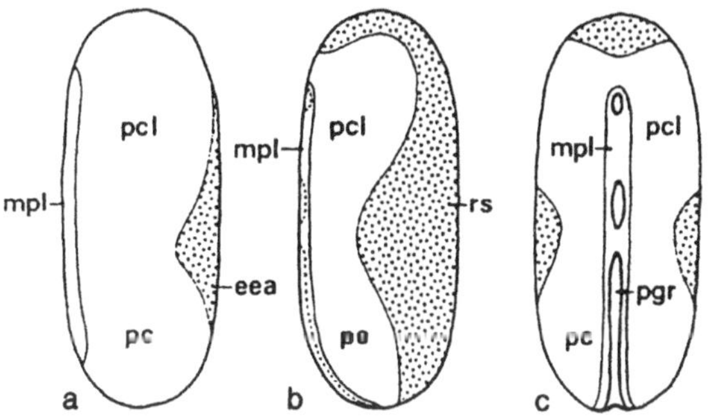


Fig. 64. Embryology. External views of embryos of *Sialis mitsuhashii* at (a) 31 hours and (b, c) 32 hours (eea, extra-embryonic area; mpl, median plate; pc, protocorm; pcl, protocephalon; pgr, primitive groove; rs, rudimental serosa) (***Suzuki*** et al. 1981).

three-segmented, and pleuropodia appear as small lobes on the first abdominal segment.

From day five the embryo increases in width and decreases in length. The antennal rudiments are three-segmented on day five and the maxillary rudiments then have three developmental points (rudiments of the lacinia, galea and palp). Thoracic legs are longer, and the neural ridges become deep. Thoracic spiracular pits develop on day six, and abdominal swellings are more evident. On days seven and eight the abdomen flexes ventrally and the head becomes distinct, leaving rudiments of the seven stemmata on each side. On day nine, the late revolution stage, the embryo expands laterally and the abdomen increases in length.

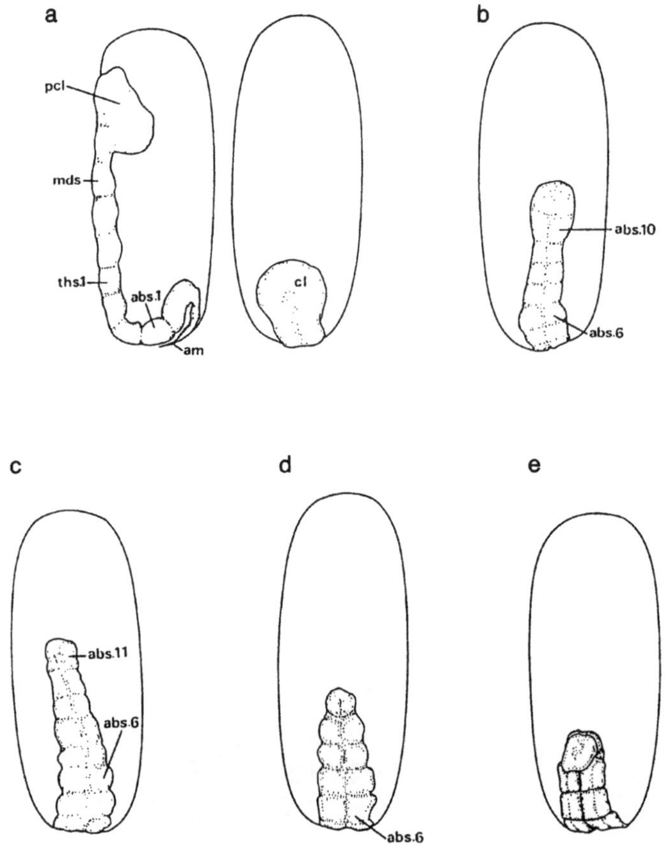


Fig. 65. Development of *Sialis matsuhashii*: a, lateral and dorsal, 2 days; b, dorsal, 3 days; c, dorsal, 4 days; d, dorsal, 5 days; e, dorsal, 6 days: pcl, protocephalic lobe; cl, caudal lobe; mds, mandibular segment; ths. 1, first thoracic segment; am, amnion; abs. 1, 6, 10, 11, abdominal segments of these numbers (***Ando*** et al. 1985).

Appendages continue to elaborate, so that the mandible rudiments bifurcate and the legs are separable into four segments representing the coxa, femur, tibia and tarsus. Elaboration continues on day ten before hatching occurs on day eleven.

The development of *Protohermes* (***Miyakawa*** 1979) is very similar to *S. mitsuhashii* in general form and sequence. A presumed primitive feature, shared between these Megaloptera, occurs at around the state of tracheal invagination (day six), when two or three swellings appear on each side of the abdominal segments. The most lateral one gives rise to the gill but others are vestigial and degenerate subsequently. Two pairs of appendage lobes, pleuropodia and the tracheal gill rudiment, occur on the first abdominal segment in *P. grandis* (***Miyakawa*** 1979, 1980). The pleuropodia in *Protohermes* are apparent from day four (Fig. 66), and ***Miyakawa*** emphasised the relevance of the two pairs of appendage rudiments in assessing homologies of abdominal appendages with the thoracic legs of insects.

Some differences occur between embryos of *Sialis* and *Protohermes* (***Ando et al.*** 1985):(1) the embryonic area in relation to the total egg surface, and that of the protocephalon in relation to the total embryo, are larger in *Sialis* at around day two; (2) cephalic lobes of *Protohermes* extend anterolaterally and do not reach the mandibular segment, whereas in *Sialis* they extend posterolaterally and reach the level of the mandibular segment; this difference is maintained in later development; (3) in late embryogenesis the outer part of the labium develops as a pair of 'comb-like' structures covering the mouthpart midline in *Sialis*, but does not develop in *Protohermes*, and (4) anal lobes in *Sialis* develop as a

Fig. 66. Development of *Protohermes grandis*: a, lateral and ventral, late 2nd day; b, ventral and lateral, early 3rd day; c, same, end 3rd day; d, same, late 4th day; e, same, 6th day; f, same, 7th day; g, lateral and ventral, 8th day; h, lateral, 9th day; i, ventral, 11th day. A1–10, 1st-10th abdominal segments; AF, abdominal gill; AL, anal lobe; AM, amnion; AAMF, anterior amniotic fold; APP, anterior pole protuberance; AT, antennal rudiment; EY, eye; IC, intercalary or premandibular segment; LB, labial segment or rudiment; LR, labral rudiment; MD, mandibular segment or rudiment; MX, maxillary segment or rudiment; PAMF, posterior amniotic fold; PCE, protocephalon; PCO, protocorm; PP, pleuropodium; RA, abdominal rudiment; SR, serosa; T1–3, thoracic segments; TL1, prothoracic leg; VCH, V-shaped line of chorion; Y, yolk (***Miyakawa***, 1979). ►

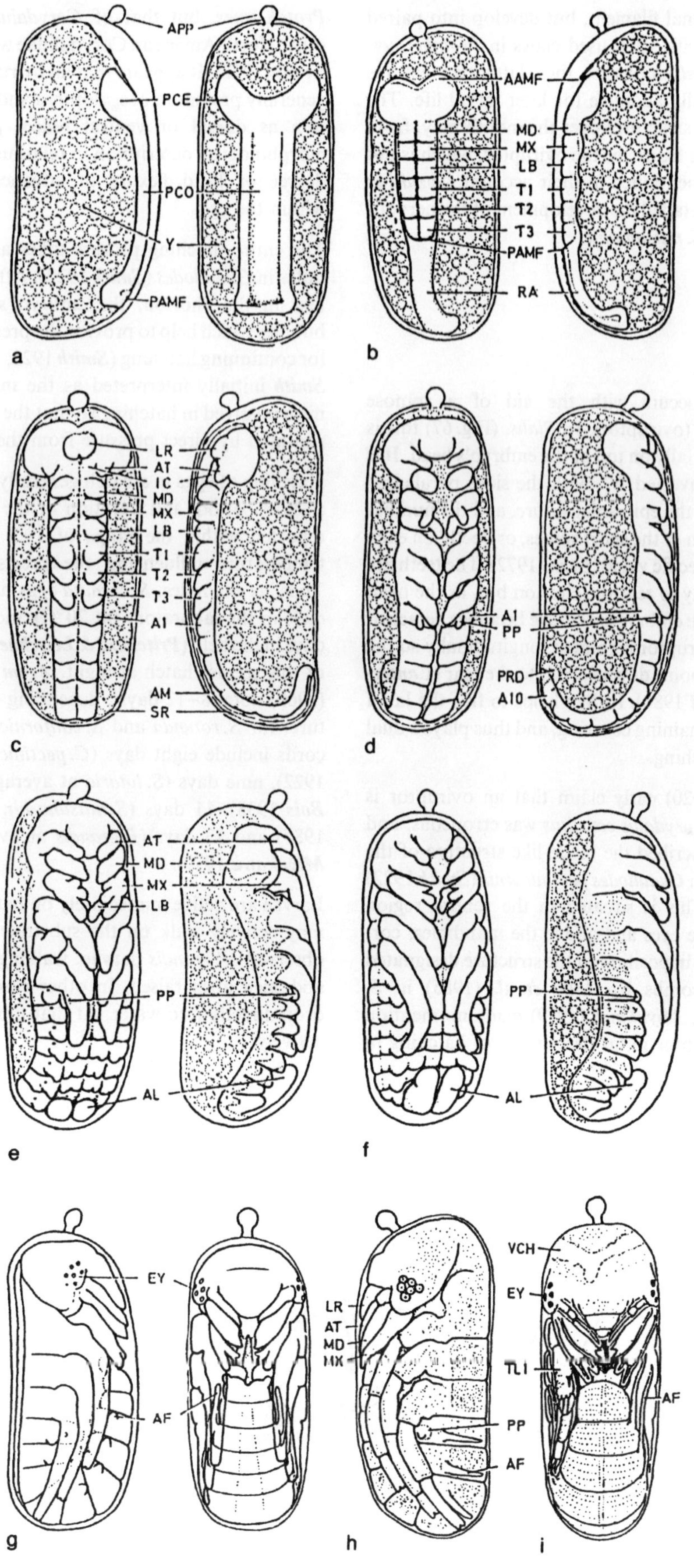
APP
PCE
PCO
Y
PAMF
a
AAMF
MD
MX
LB
T1
T2
T3
PAMF
RA
b
LR
AT
IC
MD
MX
LB
T1
T2
T3
A1
AM
SR
c
A1
PP
PRD
A10
d
AT
MD
MX
LB
PP
AL
e
PP
AL
f
EY
AF
g
LR
AT
MD
MX
PP
AF
h
VCH
EY
TL1
AF
i

single terminal filament, but develop into paired pygopods with sclerotised claws in *Protohermes*. As ***Ando et al.*** emphasise, these latter two features are essentially adaptive for later larval life. The 'comb-like' structures over the mouth may, from ***Ando et al.***'s figure, be the oviruptor, which is not otherwise mentioned in their account. ***Suzuki et al.*** (1987) detailed the development of the nervous system of *S. mitsuhashii*.

## Hatching

Hatching occurs with the aid of a spinose egg-burster (oviruptor). In *Sialis*, (Fig. 67) this is situated initially on top of the embryo's head. It is a narrow inverted 'V', with the sides paralleling the arms of the epicranial suture, and the number and position of the short spines, or 'teeth' on each arm is of specific value (***Evans*** 1972). The teeth are used initially to tear the chorion but, as the head emerges, the embryonic cuticle becomes detached and the oviruptor is rotated longitudinally so the teeth then point downward into the tear (***Canterbury & Neff*** 1980). It then helps to free the larva from its remaining covering, and thus plays a dual role in hatching.

***Smith***'s (1920) early claim that an oviruptor is absent in *Corydalus cornutus* was erroneous, and he later described the blade-like structure of the oviruptor in *Chauliodes pectinicornis* (***Smith*** 1922) (Fig. 67). This is situated in the ventral region between the eyes and above the mandibles, corresponding in position to the structure designated as labial combs by ***Ando et al.*** (1985) in *S. mitsuhashii*. ***Miyakawa*** (1979) made no mention of an oviruptor in his description of hatching in *Protohermes*, but that of *Corydalus* and some other North American Corydalidae were noted by ***Evans*** (1972). It appears that an oviruptor may be generally present in Megaloptera and its designation as dorsal or ventral reflect precision of morphological description in relation to the large degree of head flexure in advanced embryos within the eggs.

The entire hatching process takes about 45 minutes in *Chauliodes* (***Smith*** 1922). After the initial tear in the chorion, the embryo swallows air bubbles which help to provide the pressure needed for continuing hatching (***Smith*** 1920, 1922), which ***Smith*** initially interpreted as the major mechanism involved in hatching so that the chorion was ruptured by direct pressure from the embryo.

Hatching of eggs in a batch generally occurs over a short period, and duration of the egg stage is reported within the range of 1–3 weeks, with temperature influencing the precise duration. Thus, as examples, *S. cornuta* eggs hatched in 10 days in the laboratory (at 20 °C) and took 7–21 days in the field (***Pritchard & Leischner*** 1973): eggs of this species hatch at night. ***Azam & Anderson*** (1969) note '8–12 days', depending on temperature, for *S. rotunda* and *S. californica*. Other records include eight days (*C. pectinicornis*: ***Smith*** 1922), nine days (*S. lutaria* at average 19 °C: ***Du Bois*** 1938), 11 days (*S. mitsuhashii*: ***Ando et al.*** 1985) and 13 days (*P. grandis* at average 24 °C: ***Miyakawa*** 1979).

Larvae are active immediately on hatching. They may actively walk on the substrate: ***Miyakawa*** observed *P. grandis* larvae walking with their abdomen apex raised. In other cases they may drop directly into water. At that time they swim actively.

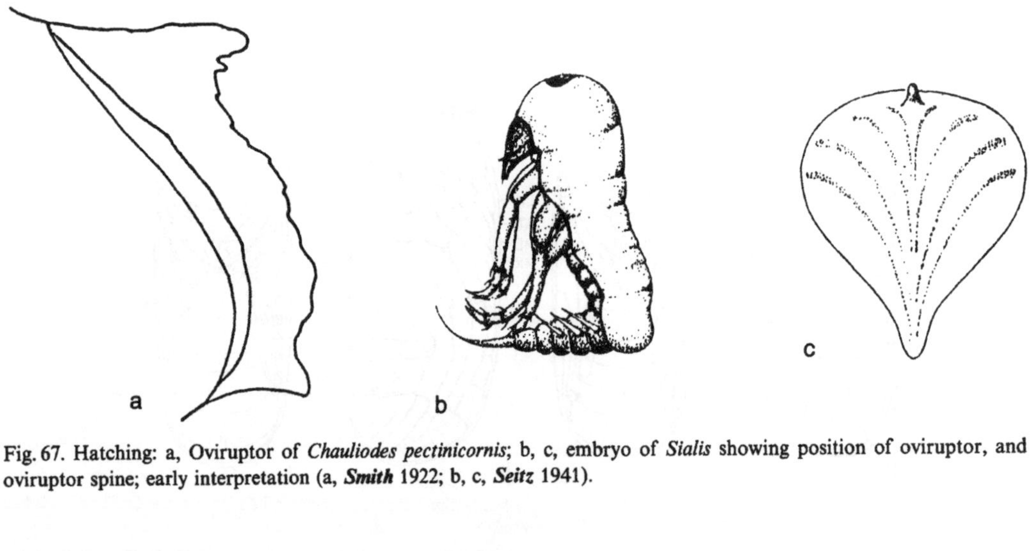


Fig. 67. Hatching: a, Oviruptor of *Chauliodes pectinicornis*; b, c, embryo of *Sialis* showing position of oviruptor, and oviruptor spine; early interpretation (a, ***Smith*** 1922; b, c, ***Seitz*** 1941).

# Larval Morphology, Anatomy and Physiology

## Morphology

The larva of each family and subfamily is very distinctive in form (see Fig. 68, and key on p. 8). Sialid larvae invariably have seven pairs of lateral abdominal gills and a terminal process, and corydalid larvae have eight pairs of gills and a pair of apical prolegs with claw-like processes. Differences between *Corydalus* and Chauliodinae are summarised by ***Cuyler*** (1958), but the major separation factor used in taxonomic keys has been the presence (*Corydalus*) or absence (Chauliodinae) of tufts of tracheal gills on the ventral side of the abdominal segments at the base of lateral gills (Fig. 74). The general appearances of the three kinds of larvae are shown in Fig. 68. All are prognathous, have a distinct head with opposable chewing mouthparts, a three-segmented thorax with three pairs of functional elongate segmented legs, and an elongate abdomen with nine or 10 distinct segments. They are, essentially, campodeiform and all instars beyond the first are grossly similar in appearance.

The appearance of first instar *Sialis* (Fig. 69), which are planktonic and positively phototactic, differs substantially from all later instars which are bottom-dwellers and negatively phototactic (***Bertrand*** 1949, ***Lestage*** 1919). *Corydalus* larvae also differ in significant features between first and later instars; *C. cornutus* first instars, for example, lack the characteristic ventral abdominal gill tufts, have no conspicuous anteclypeus, and have three-segmented antennae (***Baker & Neunzig*** 1968).

Morphology and development of *Sialis* and *Corydalus* larvae have been studied in considerable detail, but few other taxa have been appraised fully and larvae of many taxa, particularly those restricted to the Oriental Region, have not been properly characterised, although a number have been figured, and there is no reason to doubt that they are of similar form to described larvae.

There are a number of early descriptions of *Chauliodes* (s. l.) larvae and *Corydalus* from North America (examples: ***Walsh*** 1863, ***Moody*** 1877, ***Riley*** 1879, ***Weed*** 1889, ***Lintner*** 1892, ***Needham & Betten*** 1901), of varying accuracy and degree of detail. These are discussed by ***Cuyler*** (1958) and ***Evans*** (1972), who point out some of their shortcomings. The most complete morphological accounts are those of the head of *Sialis lutaria* (***Röber*** 1941) and the head and thorax of *C. cornutus* (***Kelsey*** 1954, 1957), both studies

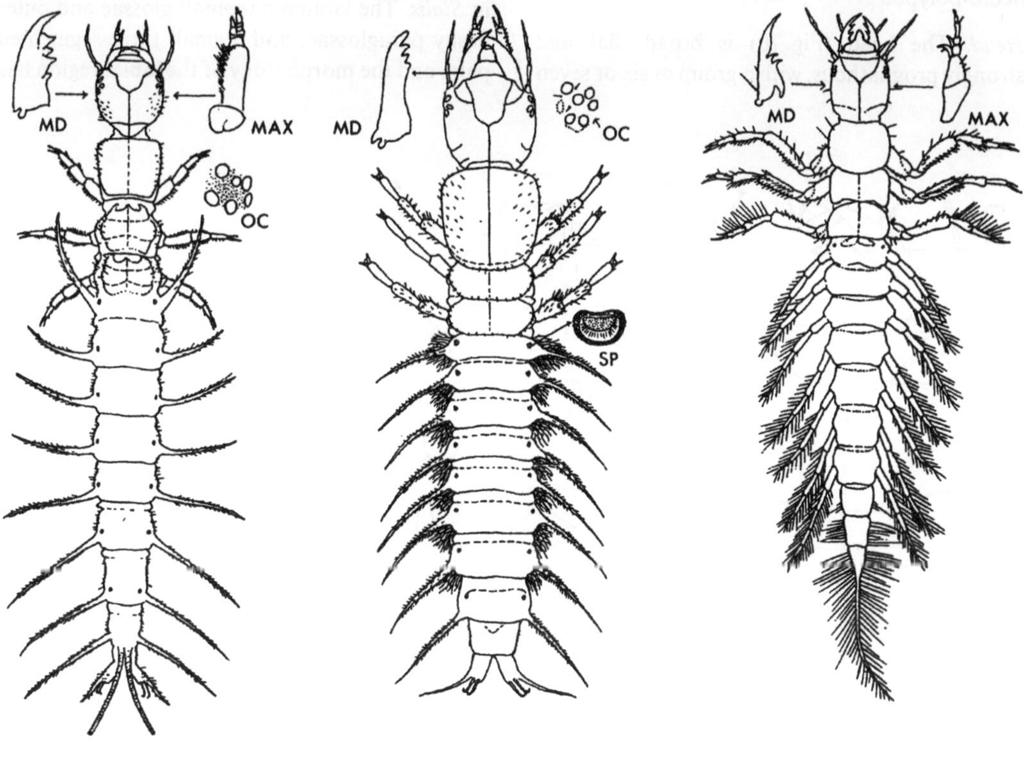


Fig. 68. Larvae: general facies of (left) *Chauliodes*, (centre) *Corydalus*, (right) *Sialis* (***Peterson*** 1960).

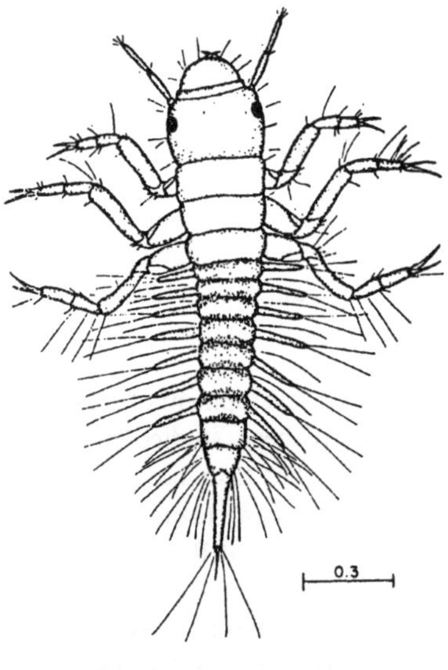


Fig. 69. First instar of *Sialis cornuta* (***Pritchard & Leischner*** 1973).

undertaken in conjunction with the corresponding adults.

Both studies are important in helping to assess homologies of structure and muscle insertions between the two stages, and the following information from these accounts refers to later instar larvae. ***Kramer*** (1955) emphasised that *Corydalus* represents one of the most primitive or generalised known kinds of endopterygote larvae, which had been referred to earlier (***Chen*** 1946), as 'campodeoid-polypod'.

*Head.* The head (Fig. 70) is broad, flat and strongly prognathous, with a group of six or seven stemmata (Fig. 71) on each side. Antennae are short, extending only about to the apex of the mouthparts and have four (*Sialis*, Chauliodinae: ***Cuyler*** 1958, ***Neunzig*** 1966) or five segments with the basal one short. The second segment bears an apical sensory process in some first instar Corydalidae (***Baker & Neunzig*** 1968). The head itself is heavily sclerotised and has a conspicuous Y-shaped epicranial suture. The stem of the Y extends about two-thirds the head length and the arms are somewhat curved or sinuous and enclose the frontal region. The clypeus is transverse and the labrum is often produced medially. In *Sialis* the pattern of crenulation each side of the extension, and the details of labral shape, can be of specific value. The labrum of *Corydalus* is narrower, with subparallel sides, and is more regularly rounded, and some other genera of Corydalidae have a broad rounded labrum. The gena is strongly angulate in *Corydalus* and *Dysmicohermes*.

Mandibles are well-developed, with several teeth, the apex usually acute and tapered. In Corydalinae there are usually three irregular teeth, and the left mandible can have an additional tooth in Chauliodinae (***Cuyler*** 1958). The maxillae are elongate. In *Corydalus* the stipes is cylindrical with a row of spines along the mesal surface, and bears a small (median) galea and (outer) four-segmented or five-segmented palp at its apex. The lacinia is not evident, but is more conspicuous in *Sialis*. The labium has small glossae and outer fleshy paraglossae, and a small three-segmented palp, and the morphology of the labial region has

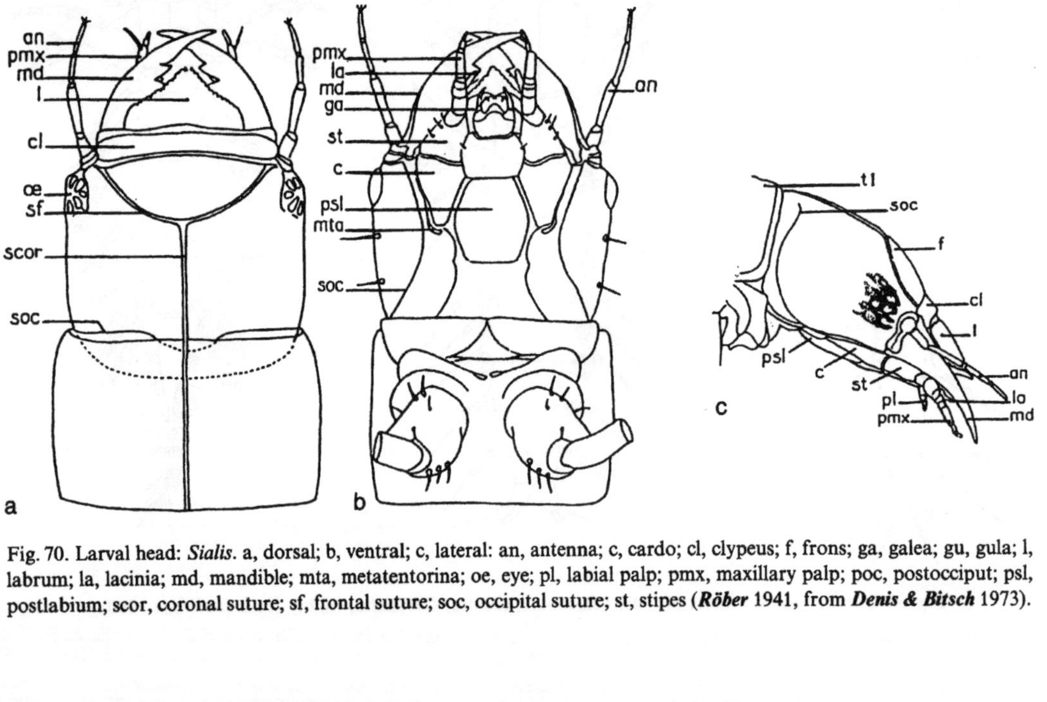


Fig. 70. Larval head: *Sialis*. a, dorsal; b, ventral; c, lateral: an, antenna; c, cardo; cl, clypeus; f, frons; ga, galea; gu, gula; l, labrum; la, lacinia; md, mandible; mta, metatentorina; oe, eye; pl, labial palp; pmx, maxillary palp; poc, postocciput; psl, postlabium; scor, coronal suture; sf, frontal suture; soc, occipital suture; st, stipes (***Röber*** 1941, from ***Denis & Bitsch*** 1973).

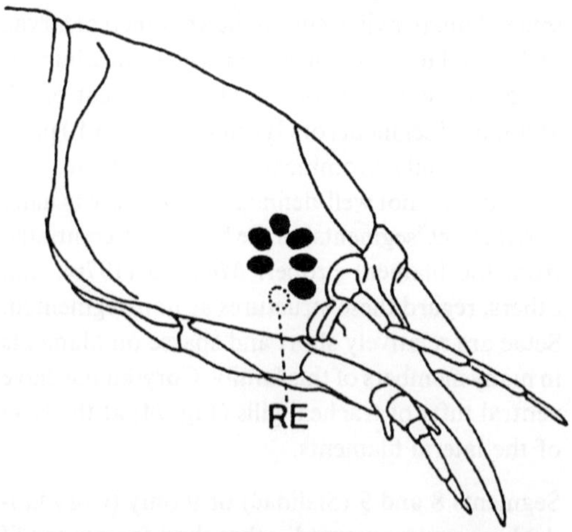


Fig. 71. Larval head of *Sialis* showing group of six stemmata and a seventh 'rudimentary eye' (RE) (***Paulus*** 1986).

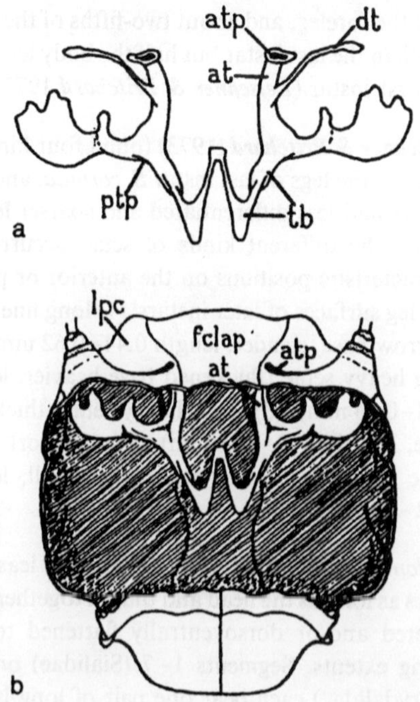


Fig. 72. Tentorium of larval *Corydalus:* a, posterodorsal; b, tentorium in head, dorsal; atp, ptp, anterior and posterior tentorial pits; at, anterior tentorial arm; dt, dorsal tentorial arm; fclap, frontoclypeal apotome; lpc, lateral postclypeus; tb, tentorial bridge (***Kramer*** 1955).

been controversial. ***Kelsey*** (1954) considered that the appearance of *Sialis* was more primitive than that of *Corydalus*, as the occipital and post-occipital sutures are widely separated, ending distally at an articulation with the mandible and maxilla, respectively. The gulamentum in *Corydalus*, defined by lateral ventral ecdysial lines and by the postgena on each side comprises the gula and mentum and is fused firmly to the remainder of the head capsule. In *Sialis* the gulamentum lies in a membranous area which allows some flexibility with the rest of the head. In *Sialis*, the postgula and lateral postgular sclerites are not separate from the gulamentum, but are distinct in *Corydalus*. Homology is confirmed by the insertion of ventral longitudinal muscles; they insert on the posterior margin of the gulamentum in *Sialis* but on the postgula of *Corydalus*. The epicranial suture is a functional ecdysial suture, as is the ventral cleavage line flanking the gulamentum (morphologically between the occipital and postocipital sutures) in *Corydalus*.

The tentorium of *Corydalus* (Fig. 72; ***Kramer*** 1955) consists largely of a pair of bars between the invaginations of the anterior and posterior tentorial pits.

*Thorax.* The prothorax (Fig. 73) is larger than either posterior thoracic segment and dorsally comprises a large quadrangular plate which extends ventrally at the sides less than in the adult. The ecdysial line continues posteriorly from the epicranial suture along nearly the whole length of the thorax, to just before the posterior margin of the metanotum. The pleural area of the prothorax is membranous dorsally, with the ventral skeletal elements similar to those in the adult. The basisternum, precoxale (precoxal bridges) and furcasternum are all heavily sclerotised (Fig. 73), and the postcoxale (postcoxal bridges) is strongly evident, though indistinct in both pupa and adult. In *Corydalus* the postnotum, laterophragma and postphragma are not as well developed in the metathorax as in the mesothorax but these segments are otherwise rather similar (***Kelsey*** 1957). The sternal area of the pterothorax is largely membranous, with the furcabases and a small central area on the basisternum being sclerotised (***Weber*** 1928, ***Kelsey*** 1957), and is broad with the leg bases widely separated. The furcasternum in *Corydalus* is delimited by the widely spaced furcabases (Fig. 73) and furcal invaginations. The poststernum is membranous. The spinasternum, present on the prothorax and mesothorax, is absent from the metathorax. Internal furcae are prominent, and the invaginated episternum is especially well-developed, in common with other neuropteroids and some Coleoptera.

All three pairs of thoracic legs are well-developed and similar in form. Each consists of coxa, trochanter, femur, tibia and tarsus (basitarsus) with a pair of simple claws. The foreleg is usually shorter than the posterior legs, and all legs of the first instar are relatively longer than in succeeding stages. In *Sialis cornuta*, for example, mid- and hind-legs are respectively 1.1 and 1.3 times longer than the foreleg, and about two-fifths of the body length in the last instar but half the body length in the first instar (***Leischner & Pritchard*** 1973).

***Leischner & Pritchard*** (1973) found four kinds of setae on the legs of last instar *S. cornuta*, and first instars had less differentiated and sparser longer setae. The different kinds of setae occurred in characteristic positions on the anterior or posterior leg surfaces of later instars: (1) long fine setae (narrow base, slender, length 0.44–0.62 mm), (2) long heavy setae (thickened base heavier, length 0.18–0.34 mm), (3) short heavy setae (thickened base, length 0.12–0.16 mm) and (4) short small setae (thickened base, dense and small, length 0.02–0.12 mm).

*Abdomen*. The abdomen is elongate, at least 1½ times as long as the head and thorax together, and tapered and/or dorsoventrally flattened to differing extents. Segments 1–7 (Sialidae) or 1–8 (Corydalidae) each bear one pair of long lateral filaments ('tracheal gills'). In Sialidae these are five-segmented (other than the first, which has four segments) and often have dense rows of long setae. Similar styli occur on the abdomen of larvae of Gyrinidae (Coleoptera), in which the abdominal apex has four hooks. Styli are present on all abdominal segments of Gyrinidae, so that 10 pairs are commonly prominent. In Corydalidae, segmentation is not well-defined but there are usually two distinct 'segments' as the base is differentiated from the filament proper. ***Matsuda*** (1976), and others, regard these structures as non-segmented. Setae are relatively short and sparse on filaments in many members of this family. Corydalinae have ventral tufts of tracheal gills (Fig. 74) at the base of the lateral filaments.

Segments 8 and 9 (Sialidae) or 9 only (Corydalidae) are unornamented, other than for groups of lateral and/or dorsal setae, often long in *Sialis*, but the apex of the abdomen differs substantially between the two families. Segment 10 in Sialidae has a long tapering median filament, which is usually strongly haired and (other than for being relatively shorter in the first instar) the segment and filament length combined is commonly at least half as long as the more anterior abdomen. Segment 10 in Corydalidae does not have such a

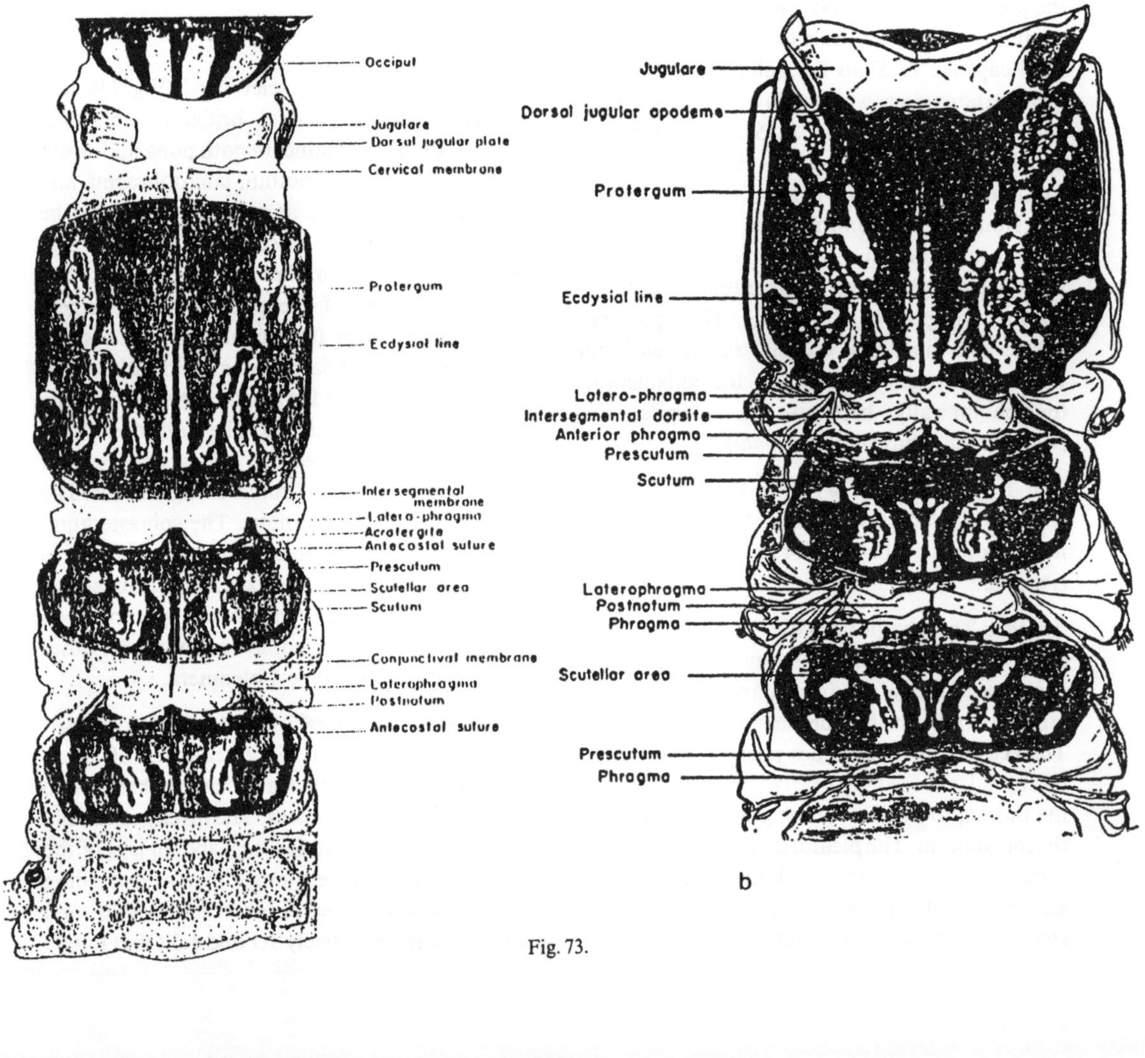


Fig. 73.

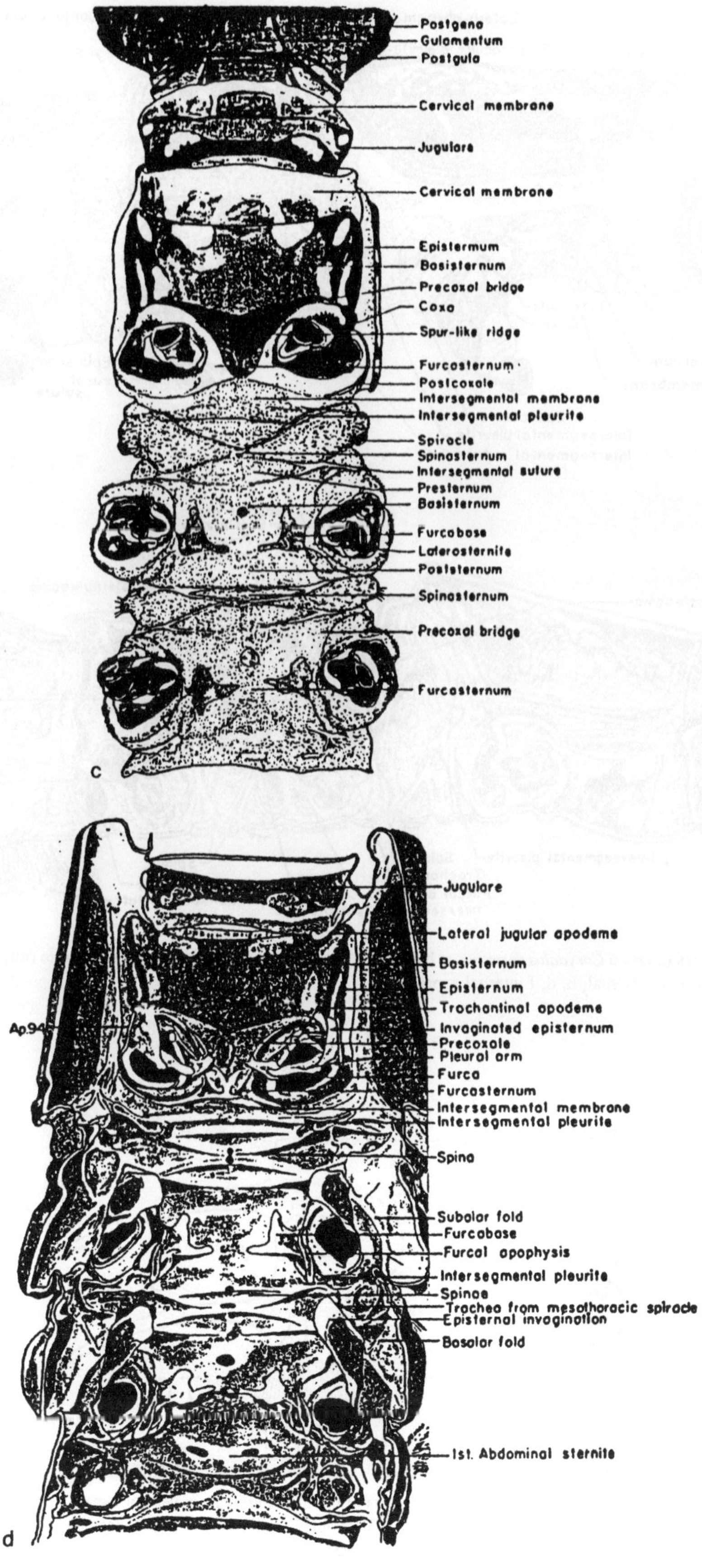

Postgena
Gulomentum
Postgula
Cervical membrane
Jugulare
Cervical membrane
Epistermum
Basisternum
Precoxal bridge
Coxa
Spur-like ridge
Furcasternum
Postcoxale
Intersegmental membrane
Intersegmental pleurite
Spiracle
Spinasternum
Intersegmental suture
Presternum
Basisternum
Furcabase
Laterosternite
Poststernum
Spinasternum
Precoxal bridge
Furcasternum
c
Jugulare
Lateral jugular apodeme
Basisternum
Episternum
Trochantinal apodeme
Ap.94
Invaginated episternum
Precoxale
Pleural arm
Furca
Furcasternum
Intersegmental membrane
Intersegmental pleurite
Spina
Subalar fold
Furcabase
Furcal apophysis
Intersegmental pleurite
Spinae
Trachea from mesothoracic spiracle
Episternal invagination
Basalar fold
1st. Abdominal sternite
d


Fig. 73.

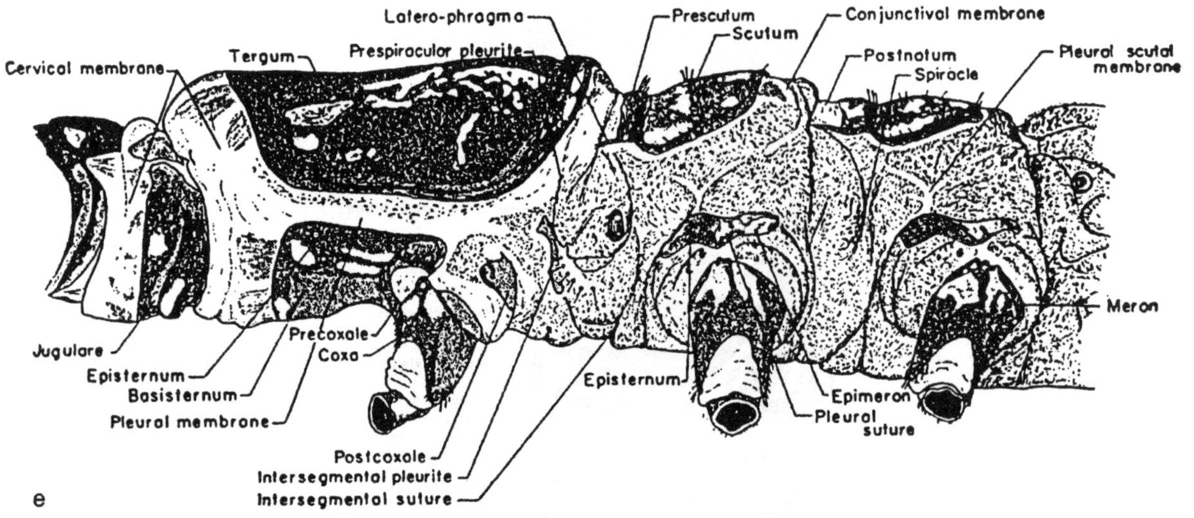


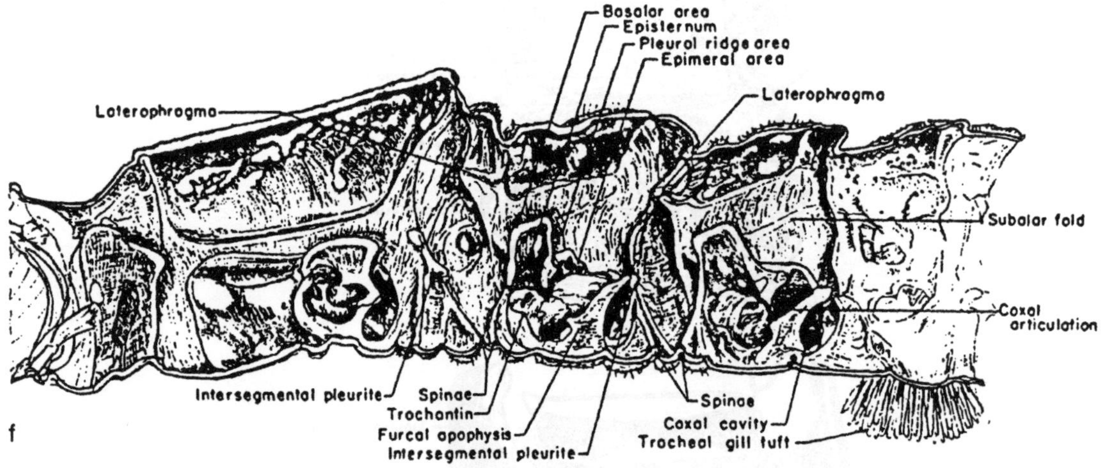


Fig. 73. Thorax of larval *Corydalus cornutus*: a, b dorsal; c, d ventral; e, f lateral (***Kelsey*** 1957). Note fully developed area of prothorax. a, c, e, external; b, d, f internal aspect.

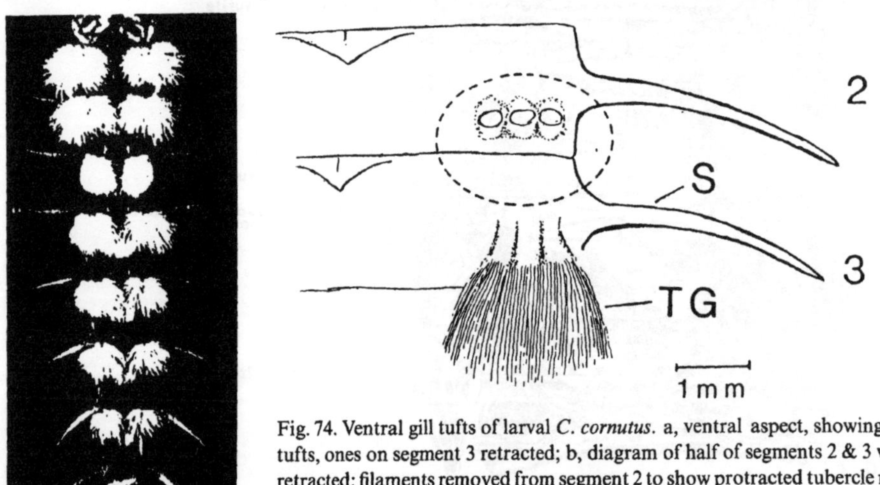


Fig. 74. Ventral gill tufts of larval *C. cornutus*. a, ventral aspect, showing seven pairs of gill tufts, ones on segment 3 retracted; b, diagram of half of segments 2 & 3 with gill (TG) on 3 retracted; filaments removed from segment 2 to show protracted tubercle near base of lateral gill (S); dotted oval indicates extent of normal protracted gill tuft (***Kinnamon & Kammer*** 1983).

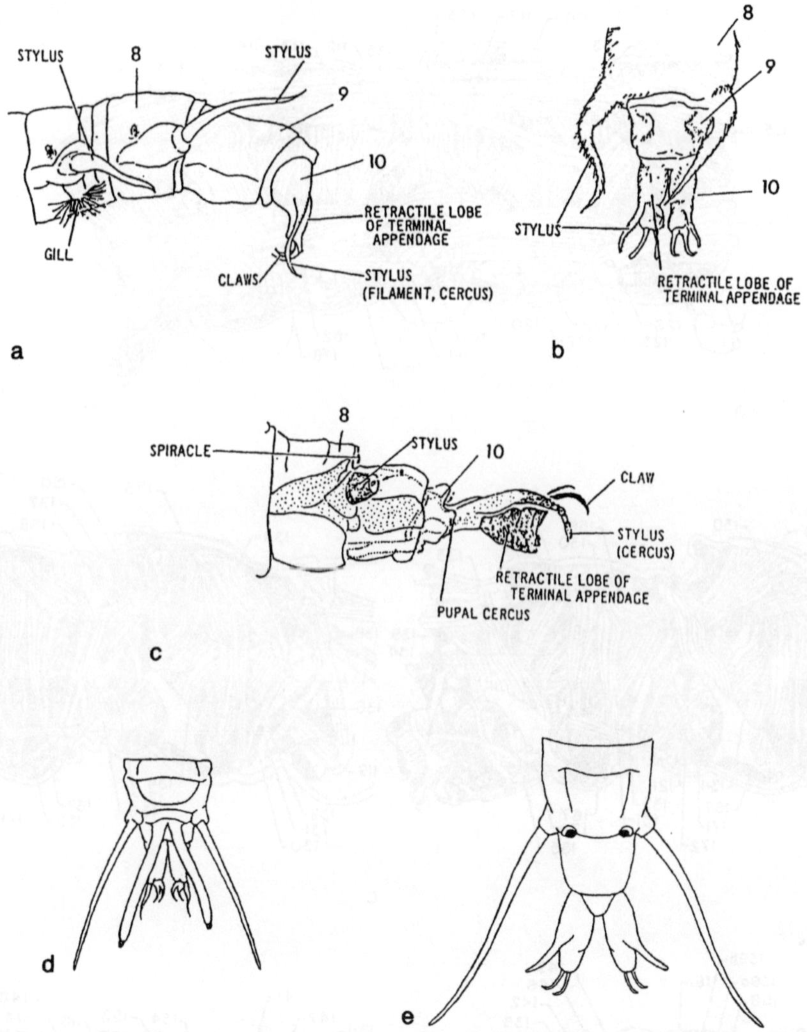


Fig. 75. Abdominal apex of larvae of Corydalidae: a, b, *Corydalus cornutus*; c, moulting male of *Neohermes californicus*; d, *Chauliodes pectinicornis*, showing extended respiratory tube; e, *N. californicus*; (a,c, lateral; b, d, e, dorsal; a, b, ***Matsuda*** 1976; c, ***Smith*** 1970; d, e, ***Cuyler*** 1958).

filament, but has a pair of large prolegs, each with an outer lateral process (lateral filament, stylus) and a pair of hooked claws. Additional ornamentation at the abdominal apex in Chauliodinae (Fig. 75) involves elaboration of the spiracles of segment 8 which, in *Chauliodes* and some other genera, are extended on long respiratory tubes (Fig. 73), which may act as 'snorkels' (***Hayashi*** 1989). The lobes bearing the claws are retractile and the outer filament of segment 10 is, following ***Matsuda*** (1976), the cercus – a homology suggested also from a study of the prepupal moult in *Neohermes californicus* (***Smith*** 1970). The filament becomes lost during the pupal stage (***Ochse*** 1944) and pupal cerci arise from the base of the larval cerci (***Smith*** 1970). The more anterior lateral filaments in *Sialis* are serially homologous with the thoracic legs (***Seitz*** 1941), and are clearly visible as embryonic appendages. They function as locomotory organs as well as in respiration (p. 73).

## Anatomy and Physiology

*Musculature*

Details of muscle arrangements of larval *Sialis* (***Seitz*** 1941, ***Röber*** 1941) and *Corydalus* (***Kelsey*** 1954, 1957, ***Kramer*** 1955) have been compared with those of the corresponding adults, and are generally similar, though no comparative study is available of the abdomen. There is no equivalent study of Chauliodinae but, from ***Maki***'s (1936) data on adult *Chauliodes*, there is little reason to doubt that similar relationships exist between the two stages.

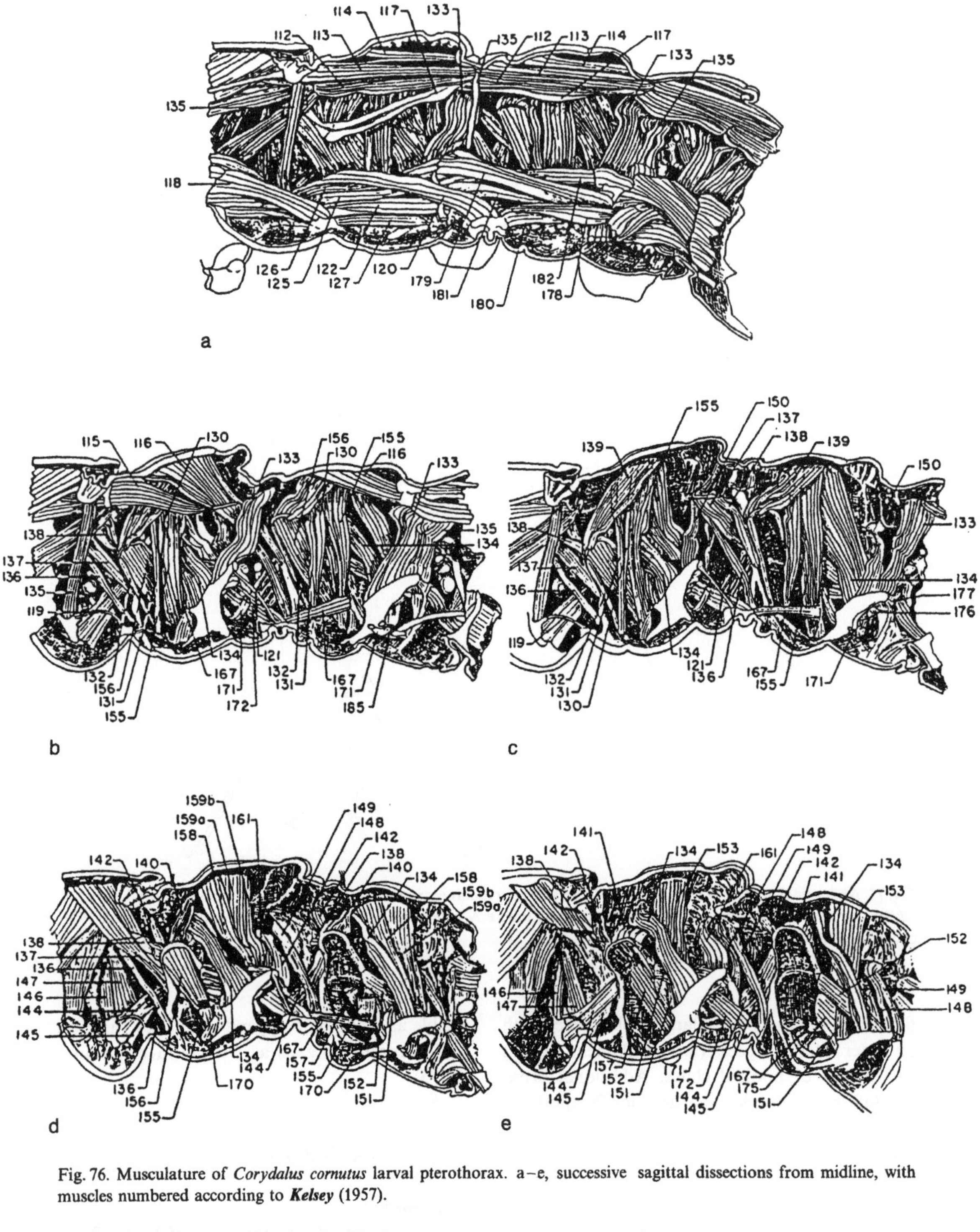


Fig. 76. Musculature of *Corydalus cornutus* larval pterothorax. a–e, successive sagittal dissections from midline, with muscles numbered according to ***Kelsey*** (1957).

In *Corydalus* antennal muscles of the larva differ from those of the adult, reflecting the position of the antennae on the anterolateral region of the head. ***Kelsey*** (1954) designates the relevant larval muscles as the antennal levator, abductor-rotator and adductor-rotator and ***Kramer*** (1955) also notes the presence of three antennal muscles. Pharyngeal muscles are better developed in the larva than in the pupa or adult and constitute three muscle groups (***Kramer*** 1955). In the pterothorax (Fig. 76) the dorsal indirect flight muscles are already present in the larva, where they contract the dorsal areas of the thorax. The mesothoracic ventral muscles are more robust in the larva, where they contract the membranous sternal region. Dorsoventral indirect flight muscles are absent in the larva, but there is a particular group of dorsoventral muscles occurring only in

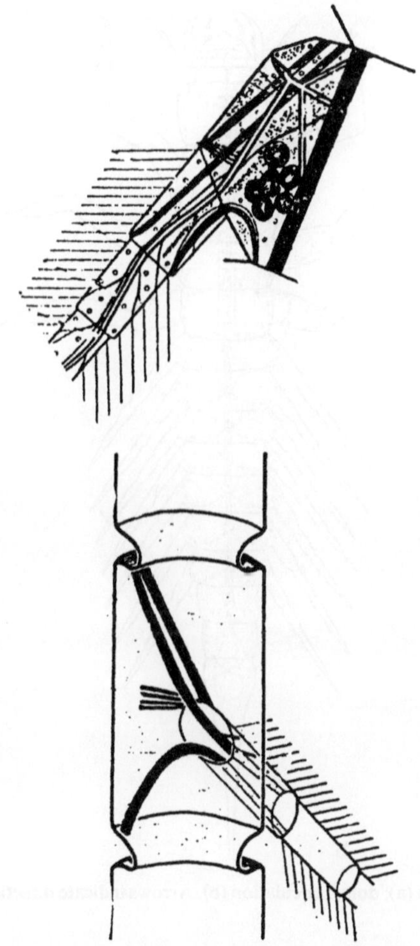


Fig. 77. Muscular articulations of gill filaments in *Sialis lutaria*: a, dorsal; b, lateral schematic from midline of segment. Muscles shown in black, schematic (after ***Ochse*** 1944).

the larva. Tergo-pleural muscles are primarily larval muscles which affect the position of the invaginated episternum and counter the pull of the strong coxal muscles. Some of the tergo-pleurals become indirect flight muscles in the adult, and others are lost. Intrapleural muscles in the larva are suspensory. At the base of the legs, the mesal muscles are smaller than in the adult and function as walking muscles.

No full description of abdominal musculature is available, but gill filaments (and the apical filament of Sialidae) are flexible and supplied with muscles (Fig. 77), which extend along the basal segment of the gill (***Ochse*** 1944).

## Circulatory system

The circulatory system of *Sialis lutaria* larvae (Fig. 78) was described in detail by ***Selman*** (1965, see also adult circulation). The dorsal vessel is a muscular tube extending from just behind the brain to abdominal segment eight. The abdominal region, the heart, extends from segments one to eight. Chambers are intersegmental, and there is a pair of segmental ostia in each of segments two to seven, which form pouches in the lateral walls of the heart and extend into the lumen to meet in the midline. The pouch walls are thus the walls of the heart chambers and the posterior walls are fused dorsally in the midline for about two-thirds of their length. Bridging with fine longitudinal muscle fibres and modification of the pouches ensures that the ostia form compound valves, through which blood passes from the haemocoel into the heart and from one chamber to another. Ostia are non-functional in abdominal segment one, where the walls are appressed to form a large valve between the first heart chamber and the thoracic aorta. The terminal chamber of the heart (segment eight) resembles the more anterior chambers. The ostia are present but their posterior walls are reduced to form a pillar of tissue between the paired internal openings, and the heart behind the final chamber is represented by two vertical sheets of connective tissue (***Selman*** 1965) containing muscle fibres attached posteriorly to the dorsal surface of the rectum.

The heart is suspended dorsally by strands of connective tissue arranged in two parallel rows and most numerous intersegmentally. Lateral and ventral walls are supported by alary muscles, of which there are two sets. The dorsal diaphragm is a nearly continuous sheet of connective tissue stretched between the alary muscles, and there are numerous pericardial cells on the dorsal surface of the diaphragm. There are also large numbers of tracheoles.

The aorta has no valves or ostia, and is a continuous tube lying between the longitudinal thoracic muscles. It twists ventrally in the head to be dorsal to the pharynx, and the dorsal wall is attached to the connective tissue sheath of the brain. The ventral wall is attached to the corpora cardiaca, so that the aorta opens ventrally and anteriorly. The aorta is highly muscular, with the wall containing radial and longitudinal muscles.

There is no ventral diaphragm and the lumen of larval appendages is not divided, but ***Selman*** (1965) noted an 'anal circulatory organ' (Fig. 79) which circulates blood in the median filament. In abdominal segments eight and nine, the lateral walls of the rectum are attached to the most lateral of the ventral longitudinal muscles, which effectively divides the abdominal apex into dorsal and ventral regions. Other muscle attachments form a

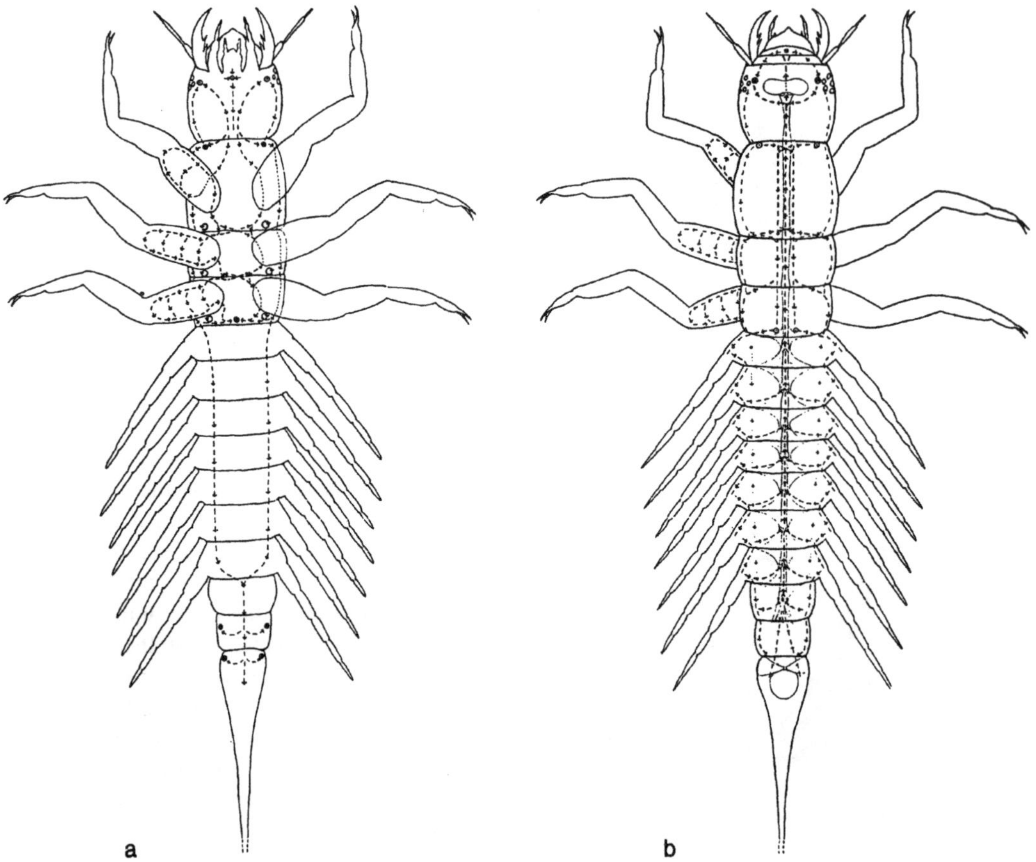


Fig. 78. Circulatory pathways of larval *Sialis lutaria*: ventral circulation (a); dorsal circulation (b). Arrows indicate direction of blood flow (***Selman*** 1965).

'funnel' chanelling blood to the filament. When the paired ventral blood streams pass into the anal circulatory organ between the internal anal dilator muscles of segment nine, contraction forces blood from the base of the median filament into the dorsal chamber, and during relaxation blood is drawn through the dilator muscles into the filament. Contraction of the anal dilator muscles is the only means by which blood is circulated in the filament and the muscles often contract without passage of faeces.

Dorsal and ventral circulatory pathways are summarised in Fig. 78. Blood enters the heart through the ostia of segments two to eight, with the amount of blood entering each pair of ostia diminishing progressively from the posterior to the anterior end. Blood is forced anteriorly through the internal heart valve into the next chamber, and the wave of contraction begins in the posterior chamber and passes forward along both heart and aorta. The subsequent wave of contraction also passes anteriorly. Peristaltic waves pass along the whole length of the aorta and, as the contractive wave passes along the mesothorax there is a considerable increase in pressure immediately anterior to the internal heart valve of the first abdominal segment, resulting in distension of the aorta. That very strong valve appears to prevent blood being forced back to the heart from aortal contraction. Blood leaving the aorta anteriorly divides into three streams – (1) an anterior stream flows downward to the bases of the mouthparts or laterally to the bases of the antennae and (2) paired lateral streams over the dorsal surface of the pharynx, which then turn forward between the brain and the mandibular muscles to between the ocelli and the antennal bases. The streams then combine ventrally and leave the head each side of the nerve commissures. There is no permanent circulation to the larval mouthparts and antennae, and blood is flushed in and out by movement of the appendages (***Selman*** 1960a, 1965).

In the thorax, movement of the legs results in blood circulation there from small blood streams between the coxal muscles, part of the main posterior lateral blood flow. Blood flow in the thorax is helped by movements of the gut and musculature: in the metathorax, particularly, blood may move very little when dorsal to the

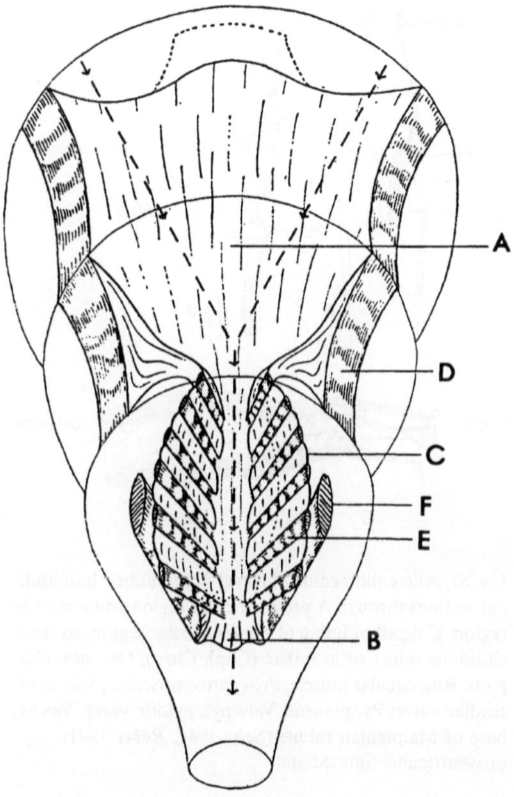


Fig. 79. *Sialis lutaria*: ventral aspect of anal circulatory organ of last instar larva, omitting ventral longitudinal muscles. Arrows indicate direction of blood flow: A, rectum; B, anus; C, connective tissue; D, outer longitudinal muscles; E, internal and F outer dilator muscles of anus (***Selman*** 1965).

aorta, and is changed there mainly by movements of the proventriculus. In the mesothorax, more blood flows between the coxal muscles and the gut than in the prothorax. Ventral blood streams divide in the posterior region of the metathorax. One pair of streams leads into the abdomen between the midline and lateral surface; the other passes dorsally from the midline and the posterior part of the foregut into the bases of the lateral gill filaments on abdominal segment one. The dorsal blood stream also divides to pass laterally to those gill bases.

In the abdomen blood flows between the fat body and the ventral longitudinal muscles, but not beside the ventral nerve cord. Blood passes into the base of each lateral filament, but part of the lateral stream bypasses the filament and follows the main tracheal trunk, and the proportion of blood doing this is influenced by the amount of fat body present in the abdomen: almost all the blood enters the gill bases in 'fat' larvae. No circulation occurs in the gills when they are at rest, and abdominal undulations by larvae help to move blood in and out of the filaments. In each of abdominal segments two to eight, part of the blood flows dorsally and into the ostia of the heart. ***Selman*** (1965) provides comprehensive details.

Blood flow in the larva is rapid, and Table 3 summarises ***Selman***'s (1965) measurements for the passage of a single blood cell through each tagma of third instar larvae at 19 °C. The average speed is about 24 cm/h, and a complete circulation takes about one minute – omitting any time spent in an appendage. Larval heartbeat rates decrease with increasing instar (3rd instar, 21–23 beats/minute to 12–14 beats/minute in the final instar – ***Selman*** 1965), possibly reflecting a general slowing of metabolism before pupation. The pupal heartbeat is intermittent.

Table 3. The time for one blood cell to circulate in various body regions of a resting 3rd instar larva of *Sialis lutaria* at 19 °C. From SELMAN, 1965.

| Region | Direction of flow | | Time (s, average of 10 records) |
|---|---|---|---|
| Head | – | – | 4.6 |
| Thorax | anterior | posterior | 21.6 |
| Abdomen | anterior | posterior | 29.0 |
| Abdomen | posterior | anterior | 4.6 |
| Thorax | posterior | anterior | 4.6 |

Duration for one complete circulation, 64.4 seconds.

## Alimentary system

The alimentary canal of *Sialis* larvae (Fig. 80) is straight and relatively simple. The narrow pharynx leads to a gradually broadened oesophagus and crop, separated by a small proventricular region from a large midgut. The hindgut is again narrowed and the rectum short and slender. There are distinct valves at each end of the midgut (Fig. 81), and the relative lengths of different parts of the canal differ between first instar and later instar larvae (***Seitz*** 1941). In the first instar the crop finishes at around the junction between prothorax and mesothorax, and the pyloric sphincter is in abdominal segment three. In later stages the crop is elongated and extends to the first abdominal segment, and the pylorus is about the level of abdominal segment six or seven. The hindgut is correspondingly shortened (Fig. 82), and the anus is near the base of the median filament.

The alimentary canal of *Corydalus* larvae has a slight forward convolution at the anterior of the

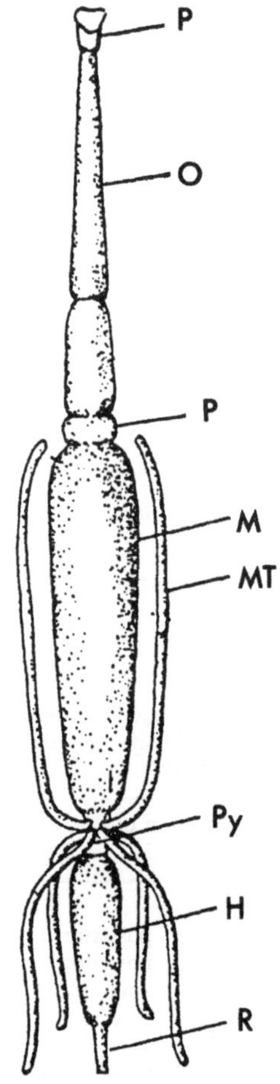


Fig. 80. Alimentary canal of larval, *Sialis*: P, pharynx (anterior); O, oesophagus; P, proventriculus (immediately before midgut); M, midgut; MT, Malpighian tubules; Py, pylorus; H, hindgut; R, rectum (***Seitz*** 1941).

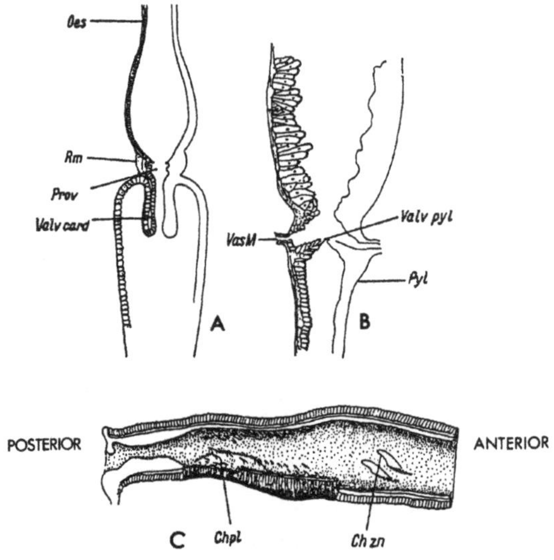


Fig. 81. Alimentary canal of larval *Sialis lutaria*, longitudinal sections through A proventricular region and B pyloric region; C detail of lining of proventricular region, to show chitinous spines or acanthae (Chpl, Chzn); Oes, oesophagous; Rm, circular muscle; Prov, proventriculus; Valv card, cardiac valve; Py, pylorus; Valv pyl, pyloric valve; Vas M, base of Malpighian tubule (***Seitz*** 1941, ***Röber*** 1941). proventriculus (immediately

hindgut, so is somewhat longer than the larval body (***Leidy*** 1848). The proventriculus is more elongate than in *Sialis*, with six rows of strong spines, and there are four cylindrical caeca towards the end of the midgut. These lie anterior to the pylorus and along the posterior region of the midgut. The hindgut gradually broadens, then narrows to the slender rectum.

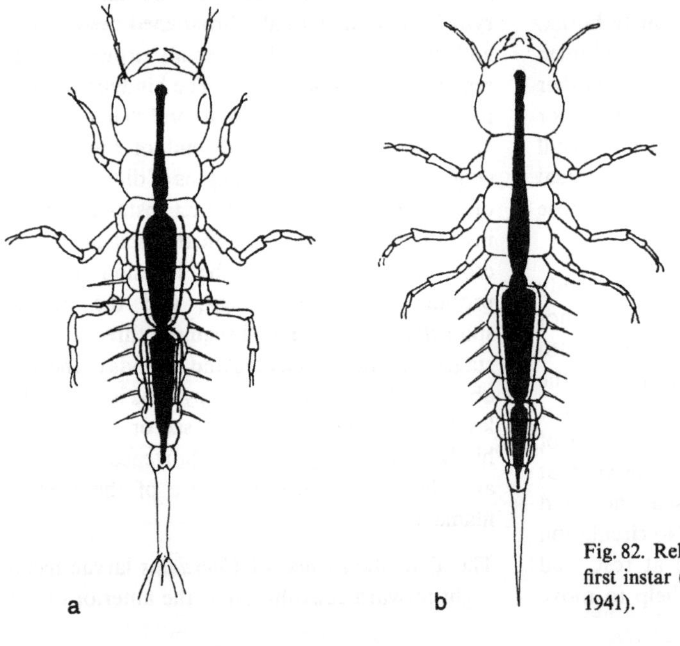


Fig. 82. Relative proportions of alimentary canal in first instar (a) and late instar (b) larval *Sialis* (***Seitz*** 1941).

Salivary glands appear to be absent in both genera.

*Sialis* larvae have six Malpighian tubules. Two are long and flexed anteriorly along the sides of the midgut, and the other four are shorter and orientated posteriorly. None is cryptonephric. *Corydalus* has eight Malpighian tubules, all long, convoluted and close to the hindgut.

The rectum is irregularly star-shaped in cross section and has well-differentiated papillae.

Ingestion by *Corydalus* larvae may not involve any chewing action but, rather, the gradual drawing of food into the mouth by alternately gripping it with the mandibles and maxillae so that small insect larvae may be found entire in the foregut (**Kramer** 1955).

## Respiratory system

Megalopteran larvae have the same number and arrangement of spiracles as adults, but these are predominantly closed and respiration occurs with the aid of the lateral and/or ventral gills and, possibly, by more general gaseous exchange over the body surface. In *Sialis lutaria*, gaseous exchange occurs most rapidly on the ventral surface of the thorax, particularly the pterothorax, and on the basal parts of the legs (**Thorpe** 1933). Larvae can live for several days after gills have been amputated. Posterior spiracles of some Corydalidae are functional. Corydalid larvae, including those which have the eighth abdominal spiracles on extended tubes, sometimes sit close to the water surface so that the spiracles are exposed to the air, especially when nearing the end of larval life. Fig. 83 exemplifies this posture in last instar *Neohermes californicus* (**Smith** 1970), in which the alternating dorsal and ventral positions of successive lateral filaments and the gripping by the terminal abdominal prolegs are characteristic. This behaviour was stimulated in the laboratory by low oxygen tension and the taxa with spiracular extensions (such as *Nigronia*) tend to inhabit still or stagnant water. Larvae of *Parachauliodes japonicus* occur in shallow water but are also 'bimodal' breathers (**Hayashi** 1989), and breath air through their caudal respiratory tube under hypoxic conditions. Their need to do this may induce them to prefer/select shallow habitats (**Hayashi & Nakane** 1989). **Smith** (1970) suggested that these modifications may indicate an early stage in the evolution of the 'breathing tube' of many aquatic Diptera and Coleoptera larvae, which represents (in part) a posterior movement of the eighth abdominal spiracles.

Although the lateral filaments are thus the primary respiratory structures in younger chauliodine larvae and Sialidae, Corydalinae have an additional series of ventral tuft-like gills. The cuticle of lateral gill filaments is very thin and each contains a large tracheal trunk with smaller branches (**Ochse** 1944, **Crass** 1949, **Smith** 1970). Corydaline ventral gills are also well-tracheated (Fig. 84) (**Barnard** 1931). They are attached to ventrolateral projections (**Snodgrass** 1935), and are ventilated by being rhythmically retracted and extended (**Kinnamon & Kammer** 1983) at rates of up to 120 times/minute. Retraction involves bending the gill's basal tubercle and filaments backward and medially towards the sternum, and extension protracts the filaments anterolaterally (Fig. 85). The gill is then held in the extended position for a short while.

Control of the ventilatory rhythm of *Corydalus*

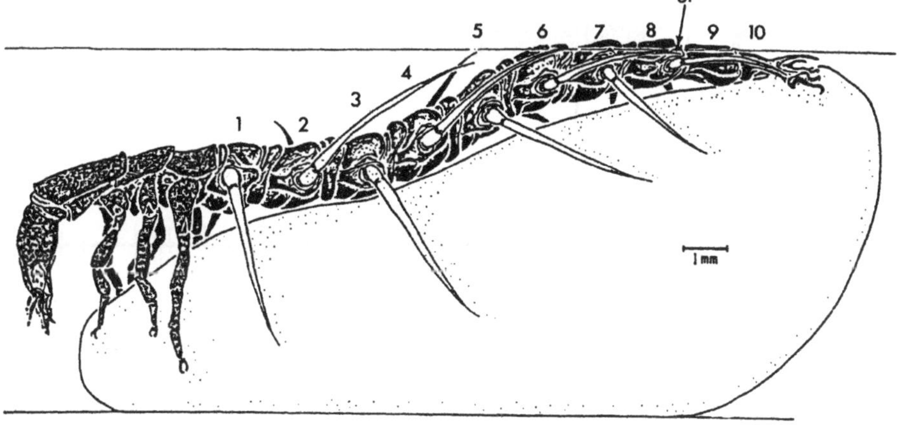


Fig. 83. Larva of *Neohermes californicus* exposing spiracles (SP) of segments 7, 8 (segments numbered) above water. Note characteristic alternating position of styli and posterior "claws" gripping the substrate (**Smith** 1970).

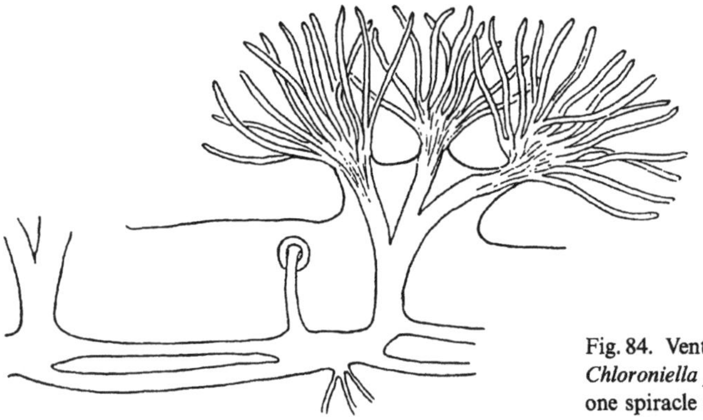

Fig. 84. Ventral gill tufts and tracheal supply of larval *Chloroniella peringueyi*. Main lateral abdominal tracheae, one spiracle and one gill tuft shown (***Barnard*** 1931).

has been studied by ***Fitch & Kammer*** (1982), ***Kinnamon & Kammer*** (1983) and ***Kinnamon et al.*** (1980, 1984). The rhythmic gill contraction typically starts with segment 3, but occasionally with segment 2 or 4. This had been noted earlier by ***Carey & Fisk*** (1965) who observed that on occasion the gills on segments 3 and 4 moved simultaneously. There is a short intersegmental delay as the activity spreads posteriorly and anteriorly, and in a strongly ventilating animal all segments participate, with the typical sequence (segment nos) 3,4,5,6,7,2,1, so that the posterior extension is much more rapid than the anterior. Only the more anterior gills are retracted in a slowly ventilating individual, and unventilating gills remain protracted (***Kinnamon & Kammer*** 1983). Ventilatory movements cease in an animal which has been laboratory-acclimated for 1–2 weeks, and the gills remain retracted. Ventilatory movements begin if such animals are disturbed or exposed to hypoxia. Frequency of ventilatory movement is increased by hypoxia (10 % $O_2$, 90 % $N_2$) but no ventilatory response is induced by hypercapnia (1–5 % $CO_2$, 20 % $O_2$, 75–79 % $N_2$) (***Kinnamon et al.*** 1984). Sensitivity to oxygen is located centrally (rather than peripherally) and is concentrated in the first abdominal ganglion in larger animals but more widely distributed amongst several abdominal ganglia in smaller individuals. Rate of gill movement also increases with water temperature, with an approximate $Q^{10}$ value of 1.88 ± 0.8 (***Carey & Fisk*** 1965). Rate of ventilatory movement is also influenced by disruption of tracheal supply (Table 4), but the frequency of retractions was not altered by removal of sensory input. Each gill retraction is caused by a burst of impulses in a single motor neuron innervating the gill retractor muscle; the neuron is in a branch of the ventral nerve (***Fitch & Kammer*** 1982).

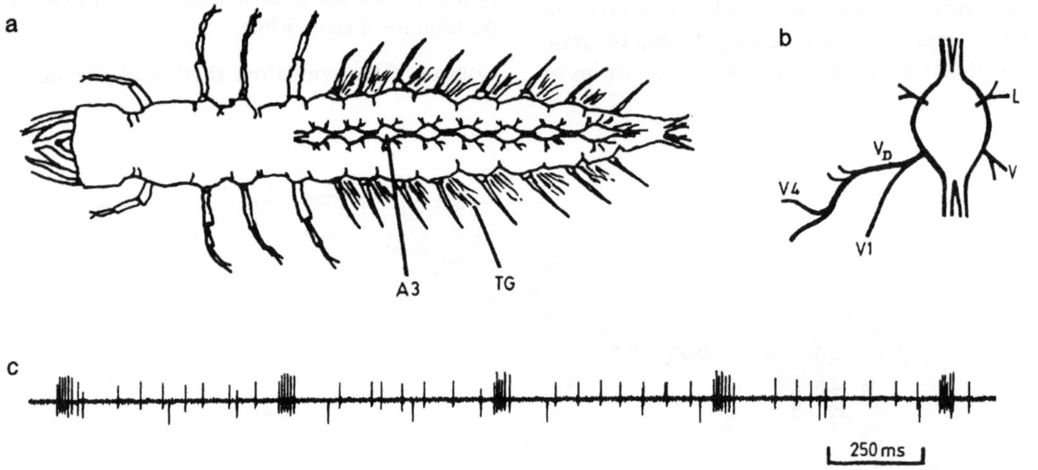


Fig. 85. *C. cornutus* larva: abdominal nervous system and gill retraction: a, dorsal dissection showing ventral nerve cord with main centre for ventilatory motor output in third abdominal ganglion (A3), and tracheal gills (TG); b, diagrammatic ventral view of one ganglion; two main nerve trunks, lateral (L) and ventral (V) occur on each side. During ventilation, a single motor neuron in V1 is responsible for gill retraction. The main distal trunk, VD, contains afferents that carry mechanosensory information from the gill tubercle; c, representativew recording from V1 showing ventilatory motor input. Each burst of impulses produces one gill retraction (***Fitch & Kammer*** 1982).

Table 4. Ventilatory responses of a representative large larva of *Corydalus cornutus* to disruption of tracheal supply. From KINNAMON et al., 1984.

| Experimental condition | Ventilating rate (gill retractions/min.) |
|---|---|
| Intact but dissected; normoxia | 36 |
| Intact but dissected; hypoxia | 66 |
| 10 min. after cutting tracheae to third thoracic ganglion; normoxia | 36 |
| 5 min. after cutting tracheal branch to gut; normoxia | 46 |
| 10 min. after above; normoxia | 82 |
| 15 min. after above; normoxia | 86 |

Muscular articulation of the *Sialis* lateral gill is shown by ***Ochse*** (1944).

Measurements of respiration (Fig. 86) on several species of Megaloptera have been related to habitats frequented, field distribution and coexistence of species. ***Evans*** (1982) postulated that a benthic species of *Sialis* (*S. rotunda*) should use less oxygen than a sympatric lotic species (*S. californica*), but was unable to confirm this from his limited data derived from Gilson respirometry.

The resting metabolism of *P. grandis* larvae varies seasonally, and is substantially lower in summer, when temperatures of the streams they frequent rise to 30 °C (Fig. 87) (***Hayashi & Yoshida*** 1987). Earlier work on *Corydalus cornutus* (***Knight & Simmons*** 1975 a, ***Brown & Fitzpatrick*** 1978) also showed adaptations to seasonal temperature variations, though *Nigronia serricornis* (which lives in summer-cold streams with maximum temperature of around 21 °C) apparently shows little such seasonal adaptation (***Knight & Simons*** 1975 b). In *Corydalus* collected from summer-

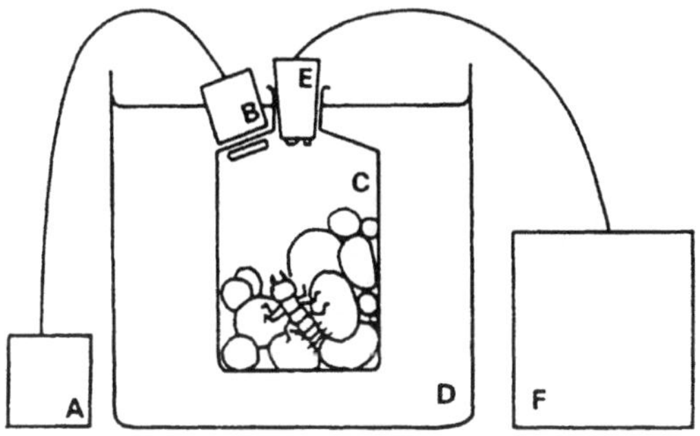


Fig. 86. Respirometer chamber used to determine oxygen consumption rate of *Protohermes grandis* larvae: A, B, magnetic stirrer; C, respiratory chamber with stones, larva, stirrer bar; D, temperature controlled waterbath; E, oxygen electrode; F, oxygen meter (***Hayashi & Yoshida*** 1987).

warm streams with a highest temperature around 31 °C the mean respiratory rate at 20 °C decreased progressively from small to medium to large larvae. Small larvae experienced the largest increase in respiration with temperature (***Knight & Simmons*** 1975 a). In spring the response of small and large larvae to temperature is the same. Respiratory response to an increase in weight is also seasonally affected. During seasonal periods of high temperature, *C. cornutus* would experience a smaller weightdependent respiratory increase than when the average temperature was close to 0 °C, allowing the larvae to be active over the whole temperature range of the habitat. *N. serricornis* larvae generally showed little effect of temperature on the relationship between weight and respiration and no such seasonal modification was therefore evident (***Knight & Simmons*** 1975 b). It thus appeared that *C. cornutus* compensates for higher summer temperatures by lowering the amount of oxygen used/unit body weight, but no such compensation occurred in *N. serricornis*. The latter appeared to be adapted physiologically so that oxygen consumption/unit body weight is relatively constant for temperature, season and stage. Cohort and individual metabolism of a univoltine population of *C. cornutus* studied by ***Brown & Fitzpatrick*** (1978), in the only such detailed study of a megalopteron, showed that metabolic compensation enabled the larvae to maintain seasonally constant respiration rates during winter and summer, with a seasonal acclimation threshold between 15 and 20 °C. Winter respiration rate was 6.07 $Jg^{-1}h^{-1}$ (5–15 °C) and summer rate was 16.15 $Jg^{-1}h^{-1}$ (20–30 °C).

*P. grandis* clearly depressed its respiratory rate at water temperatures from 20–30 °C, in common with *C. cornutus*; as ***Hayashi & Yoshida*** (1987) emphasised, the ability to acclimate to seasonal high temperatures may be especially important for such long-lived insects which experience such conditions repeatedly.

Starvation of all three species led to reduction of respiration rates. Different microhabitat distributions of *Parachauliodes japonicus* and *Protohermes grandis* may reflect their differing primary modes of respiration. If in deep water, *P. japonicus* would need to move to shore in response to decrease in oxygen concentration but could stay in deep water as long as it is sufficiently oxygenated. *P. grandis* larvae, with gill tufts, can maintain constant respiratory rates down to very low oxygen concentrations. They can minimise energy expenditure in ventilation by keeping in the better-oxygenated central 'riffles' of flowing water, as they lack the facility for aerial respiration.

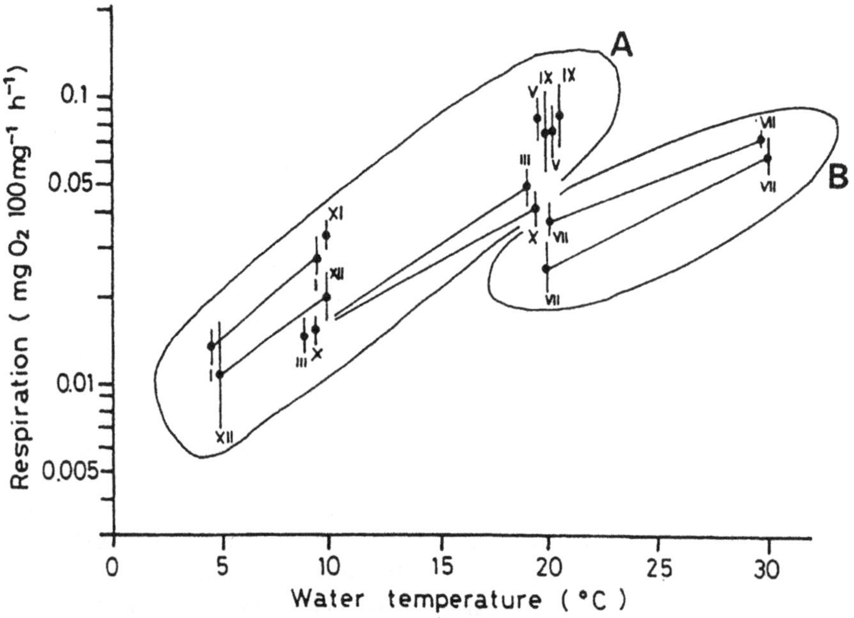


Fig. 87. Relationships between respiration rate and water temperature. Weight-specific oxygen consumption rates of larve with 100 mg dry body weight. Roman numerals are months of collection and experimentation; vertical bars are 95 per cent confidence limits. Group A are autumn, winter, spring data; group B are summer data (***Hayashi & Yoshida*** 1987: *Protohermes grandis*).

*Nigronia serricornis* in Canada is more abundant at margins of riffles, where current speeds are slower (***Fuller & Hynes*** 1987).

## Nervous system

Some details exist of the larval brain of *Sialis* (***Rahm*** 1952) and *Corydalus* (***Krauss*** 1884) but the most detailed and accurate account of the central nervous system of any megalopteron is of *Corydalus cornutus* by ***Hammar*** (1908), who emphasised that the system is rather generalised. Each thoracic segment has a distinct ventral ganglion, as do abdominal segments one to seven. The eighth abdominal ganglion is moved anteriorly to be very close to the seventh ganglion in abdominal segment seven. Thoracic and the first abdominal ganglia are beneath the large ventral muscles, but all more posterior ganglia are above them.

The brain consists externally of two large ganglia close to the roof of the head, above the oesophagus. The protocerebrum, deutocerebrum and tritocerebrum are not as prominent or well-differentiated as in the adult. The lateral borders form the optic nerve trunks and each divides into seven branches to each supply a single stemma (see also ***Yamamoto & Yoshiro*** 1975, on *Protohermes*); although there are usually only six stemmata (p. 62), the seventh is commonly present as a rudiment, and is innervated. The antennal nerve has three branches and leaves the brain anterior and ventral to the optic nerves. The clypeo-labral nerve also originates as a single trunk with the 'arched nerve' from the tritocerebrum. The circumoesophageal commissures are narrow and link the brain with the large ovoid suboesophageal ganglion, which gives rise to mandibular, maxillary, labial and salivary nerves, as well as to several smaller nerves (Fig. 88) and the median

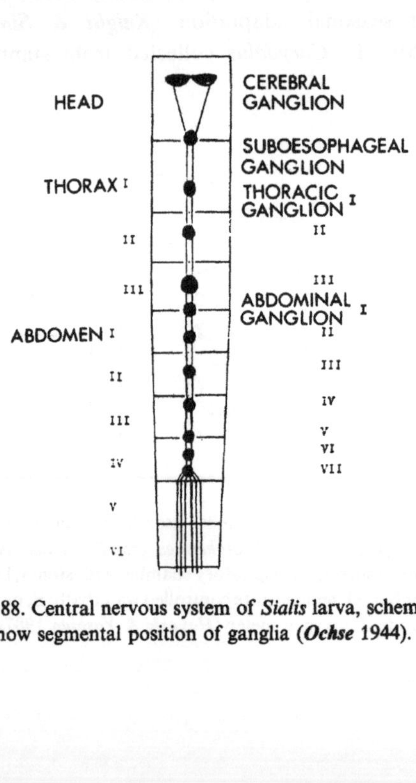


Fig. 88. Central nervous system of *Sialis* larva, schematic, to show segmental position of ganglia (***Ochse*** 1944).

nerve. The distance between the suboesophageal and prothoracic ganglion is considerably greater than between any pair of succeeding ganglia.

Three main nerve trunks arise laterally from the prothoracic ganglion. The first divides into numerous branches innervating muscles and the neck. The second innervates ventral and lateral muscles and the third goes to the foreleg. In succeeding thoracic segments the anterior nerve is small, but is enlarged to form the alar nerve in the adult, and the other two branches are homologous with the prothorax. Several nerves arise directly from the connectives (ventral nerve cords) in the thorax, but there are no such nerves in the abdomen. Nerves arising from each abdominal ganglion are rather uniform (Fig. 88) and there are lateral and ventral nerve pairs. The lateral nerve trunk innervates the more ventral muscles and controls ventilatory movements (p. 73). The eighth abdominal ganglion gives rise to the long intestinal nerves which extend forward to abdominal segment two and have a number of posterior branches to the hindgut: they represent the first (basal) branch of the lateral nerve.

Details of nerve pathways in some of the main parts of the central nervous system were described by ***Hilton*** (1911), following an earlier study (***Hilton*** 1909) on tracheation and nerves. Nearly all the main nerve trunks from the abdominal and thoracic ganglia are mixed motor and sensory, although particular pathways could be discerned and distinguished.

The 'arched nerves' anterior to the brain curve to the frontal ganglion in the midline, and the ganglion gives off a frontal nerve which bifurcates into the clypeus. ***Hammar*** (1908) described additional features of the sympathetic nervous system of *Corydalus* (Fig. 89), and ***Ehnbom*** (1948) briefly outlined some comparative aspects of Megaloptera and other neuropteroids.

## Excretion and ionic regulation

The major nitrogenous component of *Sialis* larval excretion is ammonia, which amounts to about 90% of total nitrogen (***Staddon*** 1955): *Sialis* is therefore ammonotelic, in contrast to the believed uricotelic pupa and adult. Ammonia is excreted into the hindgut, presumably via the Malpighian tubules, and hindgut contents average concentration of 136 mgN/100 ml. Liquid is expelled through the anus. Ammonia is not found in foregut fluid, and is not excreted across the general body surface. ***Staddon***'s (1955) data implied that larvae could also 'store' ammonia in some way (as ammonia did not accumulate in the tissue fluids of larvae prevented from excreting for

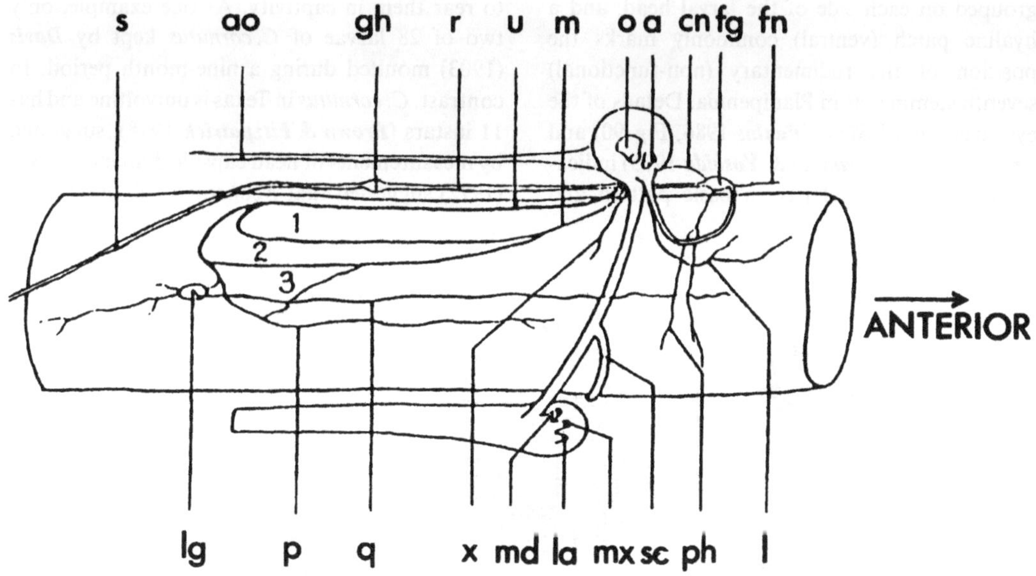


Fig. 89. Sympathetic nervous system of *Corydalus* larva: a, antennal nerve; ao, aorta; c, subpharyngeal commissure; f, frontal ganglion; fr, 'frontal nerve'; gh, ?hypocerebral ganglion; l, labral nerve; la, labial nerve; lg, corpus allatum; m, 'paracardiac nerve'with branches 1, 2, 3, p; md, mx, mandibular and maxillary nerves; nc, connecting nerve; o, optic nerve; p, nerve to lateral regions of pharynx; ph, pharyngeal nerves; q, nerve from lateral ganglia to side of the mouth; r, recurrent nerve; s, stomatogastric nerve; u, nerve partially equivalent to paracardiac and allata nerves; x, basal branch of paracardiac nerve (***Hammar*** 1908).

five days), which could possibly help to prevent accumulation of ammonia during the transition from aquatic to terrestrial life.

Further work on osmotic and ionic regulation of *Sialis* larvae was undertaken by ***Beadle & Shaw*** (1950) and ***Shaw*** (1955a, 1955b). The larvae possess no ion-absorbing mechanism, and no cuticle region is especially permeable to water (***Shaw*** 1955a), although the permeability is much greater than in terrestrial insects. The excretory fluid is not much less concentrated than the blood (***Shaw*** 1955b). Osmotic pressure of the blood is approximately equivalent to that of a 1 % sodium chloride solution (***Beadle & Shaw*** 1950). In normal *Sialis* larvae, sodium, potassium and chloride ions are absent from excretory fluid, but can be taken into the larva via the gut from the outside medium. Blood concentration is regulated by excretion of these ions via the rectal fluid (***Shaw*** 1955b). Chloride tends to be retained in the blood, but potassium and sodium are excreted more rapidly. Normal larvae do not drink water but in larvae with the blood volume reduced, water is taken up osmotically through the gut, and the gut wall is more permeable than the cuticle (***Shaw*** 1955a).

## Sense organs

Six small eucon stemmata are usually present, grouped on each side of the larval head, and a hyaline patch (ventral) commonly marks the position of the rudimentary (non-functional) seventh stemma, as in Planipennia. Details of the eye structure of *Sialis* (***Paulus*** 1986, Fig. 90) and *Protohermes* (***Yamamoto & Yoshida*** 1975) indicate considerable similarity in basic pattern and endorse earlier appraisals (***Grenacher*** 1879, ***Hesse*** 1901) of *Sialis*. There are eight (rarely nine) cone cells in *Sialis* and 10–20 in *Protohermes*. *Sialis* has about 40 retinula cells, and *Protohermes*, many more (100–300). They contain many pigment granules in *Protohermes* and are elongate, orientated perpendicular to the crystalline cone. The corneal lens is lamellated and biconvex.

### Larval instars

*Sialis* typically has 10 larval instars, the last of which leaves the water to pupate, and there can be substantial inshore migration at that time of the lifecycle. The last instar can diapause, or overwinter with little change, in Europe (***Geigy & Ochse*** 1940a, ***Rahm*** 1952, ***Elliott*** 1977, inter al.) before moving to a pupation site. However, in the laboratory, some variation can occur in instar number. Seven to 10 instars were then recorded for *S. cornuta* (***Pritchard & Leischner*** 1973), with eight or nine instars being most frequent. Individual variation in size renders instar interpretation of some field samples difficult. ***Azam & Anderson*** (1969) found 10 size categories in *S. rotunda* and *S. californica* and enumerated these as separate instars, but they were unable to determine if this number was fixed.

There are very few records or suggestions of instar number in Corydalidae, in part because their long lifecycles (p. 23) thwarted several early attempts to rear them in captivity. As one example, only two of 28 larvae of *C. cornutus* kept by ***Davis*** (1903) moulted during a nine-month period. In contrast, *C. cornutus* in Texas is univoltine and has 11 instars (***Brown & Fitzpatrick*** 1978), suggested by measurements of head capsule diameter. ***Evans*** (1972) suggested that *Neohermes californicus* has

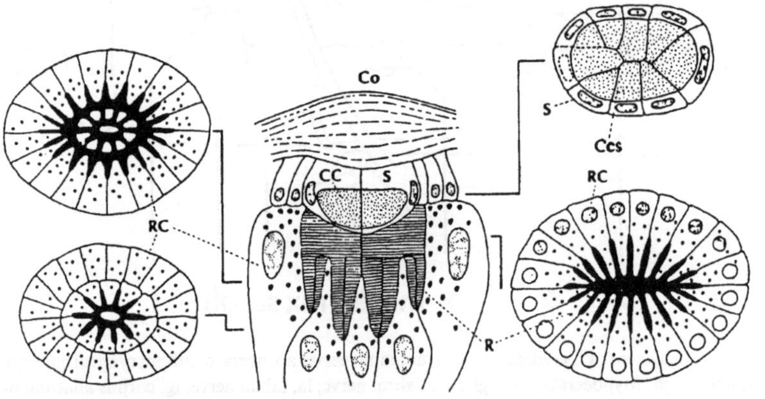


Fig. 90. Structure of stemma of *Sialis* larva: schematic longitudinal section (centre) with cross sections at levels indicated: CC, crystalline cone; Ccs, crystalline cone sector within a Semper cell; Co, cornea; R, rhabdom; Rc, retinula cell; S, Semper cell (***Paulus*** 1986).

12 instars, as he reared them through 11 instars from the egg and his largest larvae were only slightly smaller than field-collected individuals which pupated.

## Pupation and the Pupa

The exarate decticous pupae of Megaloptera are terrestrial and are typically formed in a cell in soil or under a stone or similar object near the edge of the larval habitat (Figs 91, 97).

Last instar *Sialis* commonly move some way from the water edge to pupate. ***Elliott*** (1977) cited an early British record of a larva climbing a concrete wall and passing through a vegetation thicket to pupate in a flower bed about five metres (six yards) from the water. The same transition occurs in *Corydalus*, which typically pupate a few metres or less from the water but occasionally much greater distances. ***Saunders*** (1875) noted 'as much as a hundred feet from the water's edge ... one which crawled up to the roof of a one-story building and then tumbled accidentally down the chimney' (probably the same incident as that noted by ***Riley*** 1873).

*Sialis* larvae then dig a pupal chamber 1–2 cm deep in soil and pupate in a day or so. Pupal chambers vary in form but are of three main kinds (***Seitz*** 1941, Fig. 91): (1) a closed chamber in which the pupa lies horizontally; (2) an open-mouthed vertical chamber in which the pupa lies vertically, head uppermost; and (3) a J-shaped chamber open at one end and with the pupa lying vertically in the closed arm. In captivity, *Sialis* larvae will pupate in dishes of moist soil or sand (***Azam & Anderson*** 1969, ***Evans*** 1972), and ***Pritchard & Leischner*** (1973) designed a laboratory rearing unit in which larvae were able to move directly from water to pupation cages with wet sand.

*Corydalus*, in common with at least some other Corydalidae (such as *Nigronia*: ***Cuyler*** 1965), frequently excavates a pupation chamber under a stone, rotting wood or moss (Fig. 97). Soil moisture may influence the distance of larval movement from water, as dry soil is not a good pupation substrate (***Hamilton*** 1940, ***Parfin*** 1952). In the laboratory, larvae will pupate successfully in small excavations in moist soil or sand covered with a stone (***Evans*** 1972, ***Brown & Fitzpatrick*** 1978). The pupal cell of *Protohermes* is lined with coarse gravel partially bound with silk (***Yoshida*** 1979), but no other distinct cocoon has been recorded in Megaloptera.

In some Corydalidae, the prepupal stage (Fig. 98) persists as such in the pupal chamber for three or four months (***Hamilton*** 1940) (*Archichauliodes*), and in *Corydalus* lasts for about two weeks (***Riley*** 1873).

## Metamorphosis

Control of metamorphosis (Fig. 92) in *Sialis lutaria* has been investigated in considerable detail (***Geigy*** 1942, ***Geigy & Ochse*** 1940a, 1940b, ***Geigy & Rahm*** 1951, ***Ochse*** 1944, ***Rahm*** 1952, ***Selman*** 1960b), with experiments involving chilling, ligaturing (Fig. 93) and nervous extirpation to determine the fate of various body regions and timing and sensitivity of development.

Ligaturing off the larval head can induce premature metamorphosis of the thorax and abdomen. Any instar will respond (***Selman*** 1960b) but the percentage success increases with increasing larval age. Ligaturing for five days resulted in all

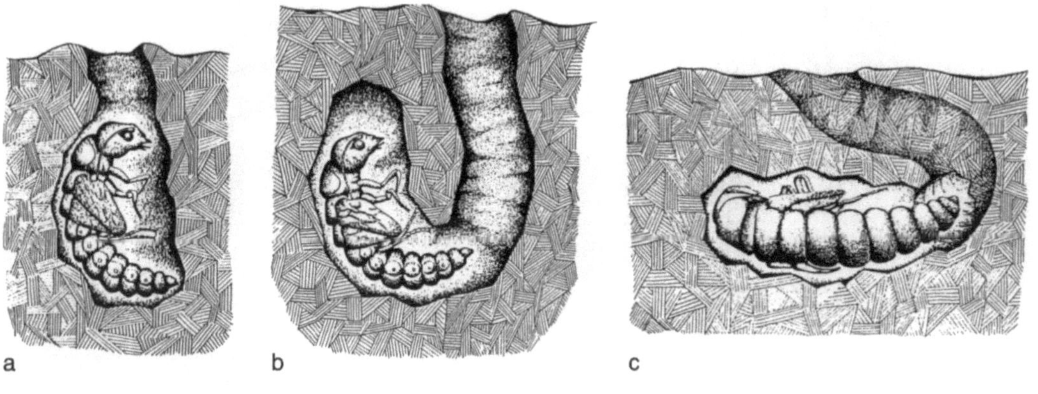


Fig. 91. Pupation cells of *Sialis*, to show variation in posture: a, normal; b, extended chamber with pupa vertical; c, extended chamber with pupa horizontal (***Seitz*** 1941). The chamber may be closed completely.

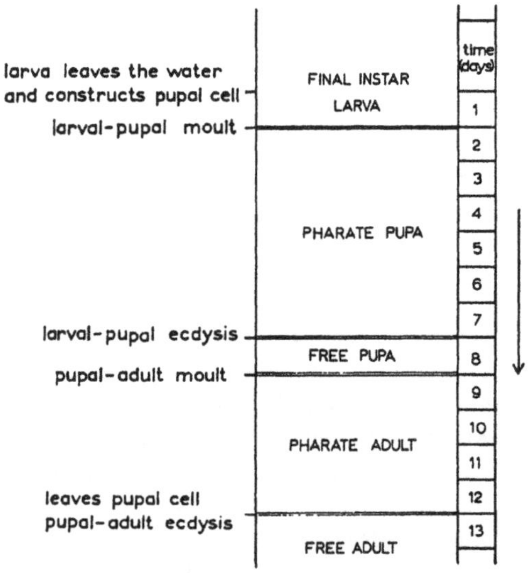


Fig. 92. Metamorphosis in *Sialis*: relationships between moults and ecdyses of *S. lutaria* (***Selman*** 1965).

moults being larval-pupal but, for shorter periods, early instars underwent larval-larval moults. Pupae from ligatured early instar larvae tended to be incompletely formed. Exposure to cold (4 °C) reduced the time taken before moulting by ligatured last instars. ***Ochse*** (1944) also investigated metamorphosis by joined (parabiotic) larvae. Occasional intermediate stage individuals have been reported: a 'larva-pupa' (larva with pupal wing pads) was illustrated by ***DuBois & Geigy*** (1935), for example (Fig. 93). Aspects of hormonal and neural control of metamorphosis were discussed, in particular, by ***Rahm*** (1952).

The pharate pupa of *Sialis* is enveloped by the last instar larval cuticle during much of its development, as the retracted gills, apical filament and ventral abdominal apex do not secrete pupal cuticle until late in the stage (***Selman*** 1960 b). Retraction of the developing pupal body from the larval cuticle, and the progressive changes in form and structure, has received considerable attention in *Sialis* and, in particular, the process and consequences of gill retraction have been described in detail. ***Selman*** (1960 b) summarised the earlier studies of ***Geigy, Ochse*** and ***Rahm*** noted above in this reexamination of the process.

Retraction of the epidermis in the terminal filament starts 24–36 hours after beginning retraction in the lateral gill filaments. Blood and epidermal cells are isolated in the old cuticle during the larval-pupal moult and are shed with that cuticle. They are discarded in the mixture of

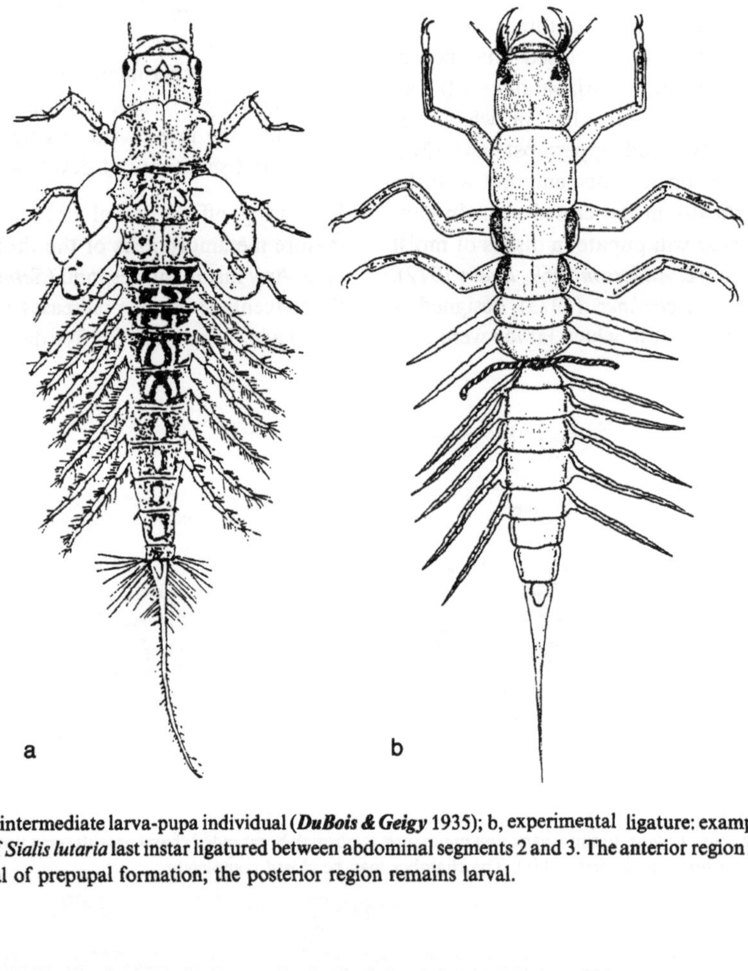


Fig. 93. a, *Sialis*, intermediate larva-pupa individual (***DuBois & Geigy*** 1935); b, experimental ligature: example from ***Geigy & Ochse*** (1940 b) of *Sialis lutaria* last instar ligatured between abdominal segments 2 and 3. The anterior region in b shows tissue retraction typical of prepupal formation; the posterior region remains larval.

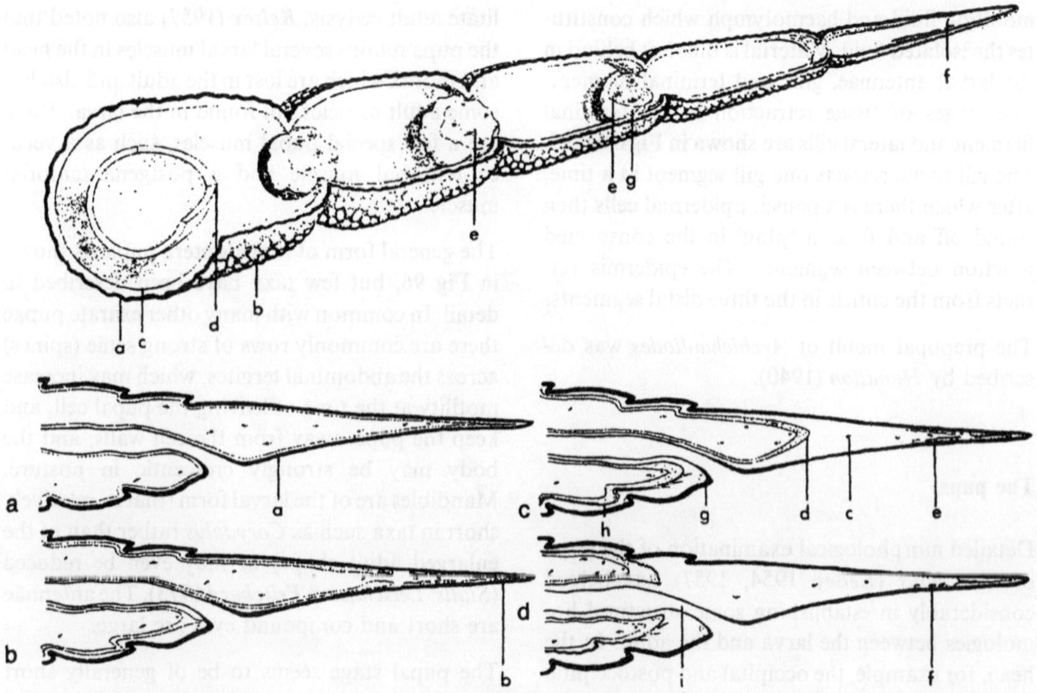


Fig. 94. **Retraction of gill and terminal filament tissue at metamorphosis in *Sialis*: Upper figure is a cut-away of whole gill after tissue retraction but before activation of moulting fluid (a, larval cuticle; b, wrinkled surface of larval cuticle; c, moulting membrane; d, tip of retracted epidermis; e, tissue plugs between segments; f, open distal end of moulting membrane; g, inner layers of moulting membrane). Below (a–d) stages of tissue retraction in the terminal filament (a, larval cuticle; b, epidermal cells isolated in tip of filament; c, blood cells isolated in fluid; d, tip of retracting epidermis; e, basement membrane of isolated epidermal cells; f, cytolysing epidermal cells; g, cells isolated from dorsal epidermis of abdomen; h, pupal cuticle; i, rectum) (*Selman* 1960a).**

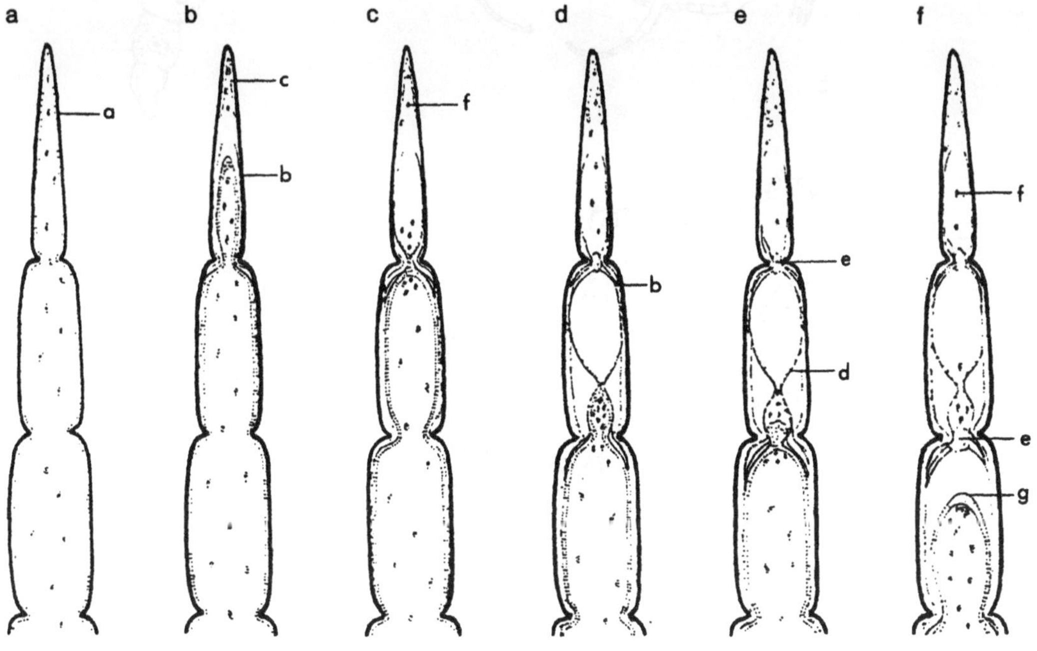


Fig. 95. Retraction of gill tissue at metamorphosis in *Sialis*, showing blood cells in fluid, seqence from a–f (a, larval cuticle; b, moulting membrane; c, epidermal cells isolated in gill tip; d, remains of cytolysed epidermal cells; e, tissue plugs between segments; f, blood cells isolated in fluid; g, tip of epidermis retracting down third gill segment) (*Selman* 1960a).

moulting fluid and haemolymph which constitutes the isolated fluid. Material is thus left behind in the larval antennae, gills and terminal filament. The stages of tissue retraction in the terminal filament and lateral gills are shown in Figs. 94, 95. The gill tissue retracts one gill segment at a time, after which there is a pause. Epidermal cells then round off and form a 'plug' in the constricted junction between segments. The epidermis retracts from the cuticle in the three distal segments.

The prepupal moult of *Archichauliodes* was described by ***Hamilton*** (1940).

## The pupa

Detailed morphological examination of the pupa of *Corydalus* (***Kelsey*** 1954, 1957) has helped considerably in establishing some structural homologies between the larva and the adult. In the head, for example, the occipital and postoccipital sutures (p. 62) are more clearly evident than in the larva, and the epicranial suture (and median ecdysial line of the thorax) are retained to facilitate adult ecdysis. ***Kelsey*** (1957) also noted that the pupa retains several larval muscles in the head and thorax which are lost in the adult and also has some adult muscles not found in the larva. There are a few special pupal muscles, such as a vertical-tentorial muscle and a postgenal-tentorial muscle.

The general form of Megaloptera pupae is shown in Fig. 96, but few taxa have been described in detail. In common with many other exarate pupae there are commonly rows of strong setae (spines) across the abdominal tergites, which may increase motility at the time of leaving the pupal cell, and keep the pupa away from the cell walls, and the body may be strongly crescentic in posture. Mandibles are of the larval form (that is, relatively short in taxa such as *Corydalus* rather than of the enlarged adult shape) or may even be reduced (*Sialis*: ***Leischner & Pritchard*** 1973). The antennae are short and compound eyes are large.

The pupal stage seems to be of generally short duration, and most published data reflect a pupal life of well under a month. For Corydalidae, *Chauliodes rastricornis*: 8–14 days, *C. pectinicor-*

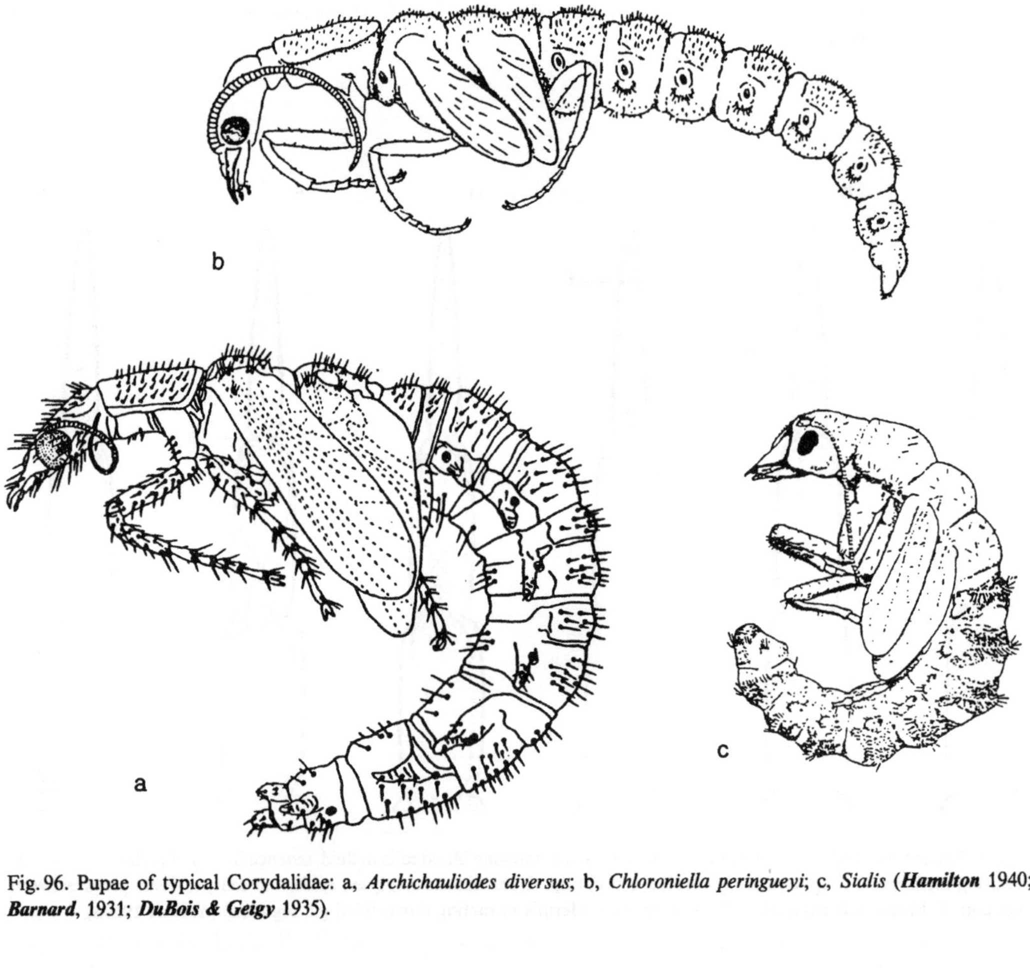


Fig. 96. Pupae of typical Corydalidae: a, *Archichauliodes diversus*; b, *Chloroniella peringueyi*; c, *Sialis* (***Hamilton*** 1940; ***Barnard***, 1931; ***DuBois & Geigy*** 1935).

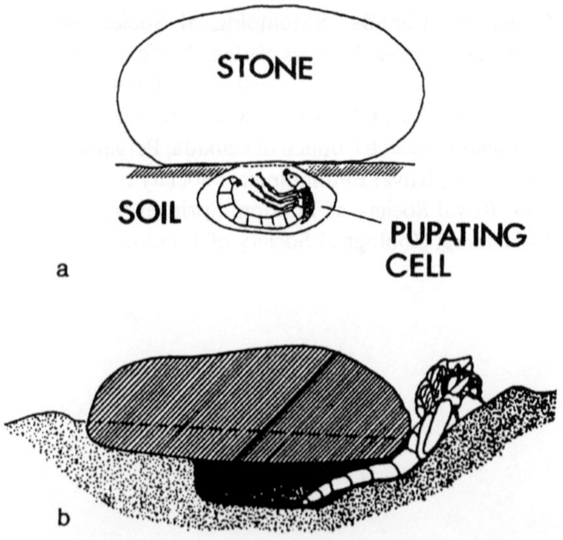


Fig. 97. Pupal cells of Corydalidae: a, *Protohermes grandis*; b, *Archichauliodes diversus*, with pupa partly emerged (***Yoshida & Hayashi*** 1985, ***Hamilton*** 1940).

*nis*: 12 days (***Needham & Betten*** 1901), *Corydalus cornutus*: 7–14 days (***Davis*** 1903), *Archichauliodes diversus*: 20–24 days, with males slightly shorter than females (***Hamilton*** 1940). *Sialis* pupae have been recorded as lasting from about a week to about a month (5–8 days: ***Azam & Anderson*** 1969; ca. 1 month: ***Pritchard & Leischner*** 1973), and ***Kaiser*** (1961) investigated the thermal requirements of two European species in some detail. *S. lutaria* needs 72.5 degree days above a threshold of 6.6 °C, so that the pupa lasts 30 days at 9.0 °C, but at 21.1 °C would take only five days. *S. fuliginosa* pupae need 84.5 degree days above 7.2 °C (relating this to 30 days at 10 °C, 20 days at 11.4 °C, 10 days at 15.7 °C and five days at 24.1 °C). Some published variations in pupal duration are thus likely to reflect differing temperature regimes. The pupa of *A. diversus* is negatively phototropic and will burrow if exposed to light (***Hamilton*** 1940).

If the pupal cell is open (as in some *Sialis* burrows), the final moult occurs in the cell and the cast pupal skin is left there. If the cell is closed the pupa itself digs out and emerges to moult on the surface. Corydalid pupae may escape by digging tunnels adjoining the surface of the covering stone (Fig. 97) and moult without emerging completely from the tunnel (***Hamilton*** 1940). The newly-moulted adult remains more or less still until wings have expanded and the body has hardened. Meconial fluid is passed.

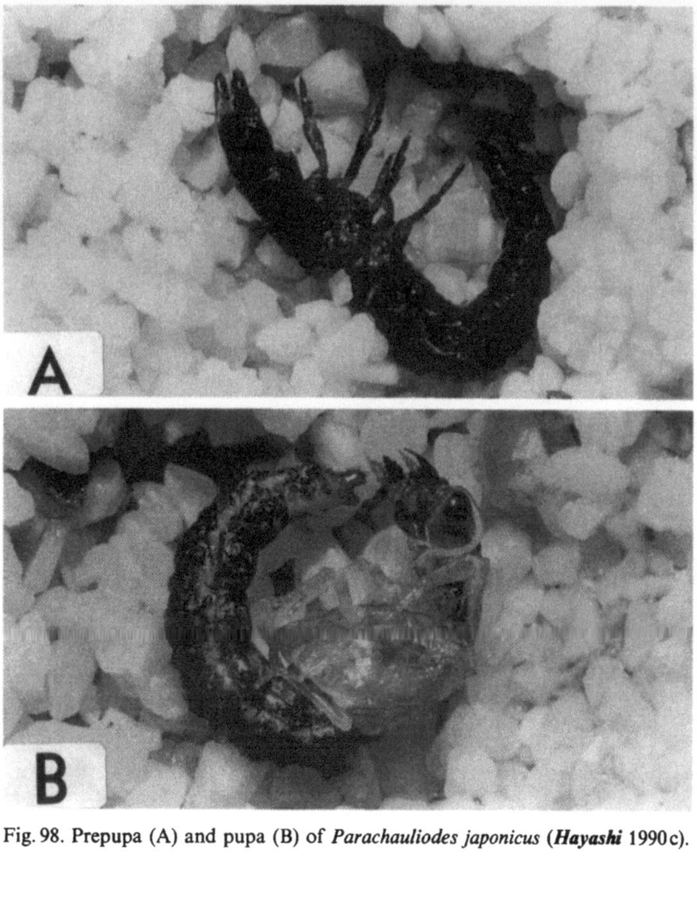


Fig. 98. Prepupa (A) and pupa (B) of *Parachauliodes japonicus* (***Hayashi*** 1990c).

*Acknowledgments*

We are very grateful to Dr. Oliver S. Flint, Jr. for his perceptive comments on a draft of this account.

The following publishers are thanked for permission to reproduce or modify figures and/or tables: Ecological Society of America; Entomological Society of Canada; Entomological Society of America; Goecke & Evers Verlag, Krefeld; Kluwer Academic Publishers; Melbourne University Press; The Natural History Museum, London; National Research Council of Canada; Pergamon Press PLC; Royal Entomological Society of London; Royal Society of London; Springer Verlag, Heidelberg; Zoological Society of London.

# Literature cited


***Achtelig, M.*** (1967): Über die Anatomie des Kopfes von *Raphidia flavipes* Stein und Verwandtschaftsbeziehungen der Raphidiidae zu den Megaloptera. – Zool. Jahrb. Anat., Jena, 84: 249–312.

– (1975): Die Abdomenbasis der Neuropteroidea (Insecta, Holometabola). Eine vergleichend anatomische Untersuchung des Skeletts und der Muskulatur. – Zoomorphologie, Berlin, 82: 201–242.

– (1976): Indizien zur Monophylie der Raphidioptera und Megaloptera (Insecta, Holometabola). – Verh. Dtsch. Zool. Ges., Stuttgart, 1976: 233.

– (1978): Entwicklung und Morphologie der inneren und äußeren weiblichen Genitalorgane der Kamelhalsfliegen (Neuropteroidea: Raphidioptera). – Ent. Ger., Stuttgart, 4: 140–163.

– (1981): Revisionary notes. In ***Hennig, W.*** 1981. – Insect Phylogeny. John Wiley, Chichester, 514 pp.

***Achtelig, M. & Kristensen, N.P.*** (1973): A re-examination of the relationships of the Raphidioptera (Insecta). – Z. zool. Syst. Evol.-forsch., Frankfurt/Main, 11: 268–274.

***Acker, T.S.*** (1960): The comparative morphology of the male terminalia of Neuroptera. – Microentomology, Stanford, Calif., 24: 25–84.

***Afzelius, B.A. & Dallai, R.*** (1988): Spermatozoa of Megaloptera and Raphidioptera (Insecta, Neuropteroidea). – J. ultrastr. mol. Struct. Res. 101: 185–191.

***Albarda, H.*** (1891): Revision des Rhaphidides. – Tidskr. Ent., Stockholm, 34: 65–184.

***Ando, H., Miyakawa, K. & Shimuzu, S.*** (1985): External features of *Sialis mitsuhashii* embryo through development. In: ***Ando, H. & Miya, K.*** (eds.). – Recent Advances in Insect Embryology in Japan, pp. 191–201. ISEBU, Tsukaba.

***Arnold, J.W.*** (1964): Blood circulation in insect wings. – Mem. ent. Soc., Can., Ottawa, 38, 60 pp.

***Asahina, S.*** (1987): A new corydalid fly from Okinawa Island. – Gekkan-Mushi, Tokyo No. 201: 17–19.

***Aspöck, H. & Aspöck, U.*** (1971): Raphidioptera (Kamelhalsfliegen). – Handbuch der Zoologie, 4 (2) 25: 1–50.

***Aspöck, H., Aspöck, U. & Hölzel, H.*** (1980): Die Neuropteren Europas. – Goecke & Evers, Krefeld: I, 495 pp; II, 355 pp.

***Aspöck, U. & Aspöck, H.*** (1983): *Sialis vanderweelei* n. sp.: Erstnachweis der Familie Sialidae in Nordafrika (Neuropteroidea: Megaloptera). – Ent. Z. Frankfurt, 93: 17–20.

***Ast, F.*** (1920): Über den feineren Bau der Facettenaugen bei Neuropteren. – Zool. Jahrb. Anat., Jena, 41: 411–458.

***Azam, K.M. & Anderson, N.H.*** (1969): Life history and habits of *Sialis rotunda* and *S. californica* in western Oregon. – Ann. ent. Soc. Amer., Lanham, Maryland 62: 549–558.

***Baker, J.R. & Neunzig, H.H.*** (1968): The egg masses, eggs and first-instar larvae of eastern North American Corydalidae. – Ann. ent. Soc. Amer., Lanham, Maryland, 61: 1181–1187.

***Balduf, W.V.*** (1939): The bionomics of entomophagous insects. Part 2. – Swift, New York.

***Banks, N.*** (1940): Report on certain groups of neuropteroid insects from Szechwan, China. – Proc. US. nat. Mus., Washington, 88: 173–220.

– (1943): Neuroptera of northern South America. – Bol. Ent. Venez., 2: 59–66.

***Barnard, K.H.*** (1931): The Cape Alder-Flies (Neuroptera, Megaloptera). – Trans. Roy. Soc. S. Africa, Cape Town, 169–184.

– (1940). Additional records, and descriptions of new species, of South African Alder-flies (Megaloptera), May-flies (Ephemeroptera), Caddis-flies (Trichoptera), Stone-flies (Perlaria), and Dragon-flies (Odonata). – Ann. Sth. Afr. Mus., Cape Town, 32: 609–615.

***Beadle, L.C. & Shaw, J.*** (1950): The retention of salt and the regulation of the non-protein nitrogen fraction in the blood of the aquatic larva *Sialis lutaria*. – J. exp. Biol., Edinburgh, 27: 97–109.

***Benedict, P.R. & Fisher, G.T.*** (1972): Commensalistic relationships between *Plecopteracoluthus downesi* (Diptera: Chironomidae) and *Chauliodes pectinicornis* (Megaloptera: Corydalidae). – Ann. ent. Soc. Amer., Lanham, Maryland, 65: 109–111.

***Berland, L. & Grassé, P.P.*** (1951): Ordre des Megaloptères. pp. 5–17. In: ***Grassé, P.P.*** – (ed.) Traité de Zoologie, Masson, Paris,

***Bertrand, H.*** (1949): Notes morphologiques sur les larves des *Sialis L.* (Megalopteres, Sialidae). – Feuille Nat., Paris, 4: 5–10.

***Bitsch, J.*** (1979): Morphologie abdominale des Insectes pp. 291–578 In: ***Grassé, P.*** – (ed.) Traité de Zoologie, Masson, Paris, 8 (2).

***Bonhag, P.F.*** (1958): Ovarian structure and vitellogenesis in insects. – Ann. Rev. Ent., Palo Alto, Calif., 3: 137–160.

***Boudreaux, H.B.*** (1979): Arthropod phylogeny with special reference to insects. – Wiley-Interscience. New York.

– (1980): Proventricular acanthae and their phylogenetic implications. – Ann. ent. Soc. Amer. Lanham, Maryland, 73: 189–196.

***Bowles, D.E.*** (1990): Life history and variability of secondary production estimates for *Corydalus cornutus* (Megaloptera: Corydalidae) in an Ozark stream. – J. Agric. Entomol., Clemson, S.C. 7: 61–70.

***Brigham, W.U.*** (1982): Megaloptera. In ***Brigham, A.R., Brigham, W.U. & Gnilka, A.*** – (eds) Aquatic insects and oligochaetes of North and South Carolina pp. 7.1–7.12. Midwest Aquatic Enterprises, Mahomet, Illinois.

***Brooker, M.P.*** (1979): The life cycle and growth of *Sialis lutaria* L. (Megaloptera) in a drainage channel under different methods of plant management. – Ecol. Entomol., Oxford, 4: 111–117.

***Brown, A.V. & Fitzpatrick, L.C.*** (1978): Life history and population energetics of the dobson fly, *Corydalus cornutus*. – Ecology, Tempe, Arizona, 59: 1091–1108.

***Büning, J.*** (1979): The telotrophic-meroistic ovary of Megaloptera I. The ontogenetic development. – J. Morph., Philadelphia, 162: 37–66.

– (1980): The ovary of *Raphidia flavipes* is telotrophic and of the *Sialis* type. (Insecta, Raphidioptera). – Zoomorphol., Berlin, 95: 127–131.

***Burmeister, H.*** (1832): Handbuch der Entomologie. I. Berlin.

– (1839): Handbuch der Entomologie. II. Berlin.

***Butler, R.L., Copper, E.L., Crawford, J.K., Hales, D.C., Kimmel, W.G. & Wagner, C.E.*** (1973): Fish and food organisms in acid mine waters of Pennsylvania. – U.S. Environ. Prot. Agency, Washington, Ecol. Res. Series No R3-73-032, 158 pp.

***Caldwell, B.A.*** (1976): The distribution of *Nigronia serricornis* and *Nigronia fasciatus* in Georgia and water chemistry parameters associated with the larvae (Megaloptera: Corydalidae). – Bull. Georgia Acad. Sci., 34: 24–30.

***Canterbury, L.E.*** (1978): Studies on the genus *Sialis* (Sialidae: Megaloptera) in eastern North America. – Ph.D. Thesis, University of Louisville, Kentucky.

***Canterbury, L.E. & Neff, S.E.*** (1980): Eggs of *Sialis* (Sialidae: Megaloptera) in eastern North America. – Can. Ent., Ottawa, 112: 409–419.

***Carey, W.E. & Fisk, F.W.*** (1965): The effect of water temperature and dissolved oxygen content on the rate of gill movement of the hellgrammite *Corydalus cornutus*. – Ohio, J. Sci., 65: 137–141.

***Carpenter, F.M.*** (1954): Fossil orders. In ***Brues, C.F., Melander, A.L. & Carpenter, F.M.*** Classification of Insects. – Bull. Mus. Comp. Zool., Harvard, Cambridge, Mass, 108: 777–827.

***Cazal, P.*** (1948): Les glandes endocrines retrocerebrales des Insectes (étude morphologique). – Bull. biol. Fr. Belg., Suppl, Paris, 32: 1–127.

***Chandler, H.P.*** (1954): Four new species of dobsonflies from California. (Megaloptera: Corydalidae). – Pan.-Pac. Entomol., San Francisco, 30: 105–111.

– (1956): Megaloptera. In **Usinger, R.L.** – (ed.) Aquatic Insects of California pp 229–233. University of California Press, Berkeley.

***Chen, S.H.*** (1946): Evolution of the insect larva. – Trans. Roy. Ent. Soc. Lond., 97: 381–403.

***Coope, G.R.*** (1961): On the study of glacial and interglacial insect faunas. – Proc. Linn. Soc. Lond., 172: 62–65.

***Crampton, G.C.*** (1918): The genitalia and terminal abdominal structures of male Neuroptera and Mecoptera with notes on the Psocidae, Diptera and Trichoptera. – Psyche, Cambridge, Mass., 25: 47–58.

– (1926): A comparison of the neck and prothoracic sclerites throughout the orders of insects from the standpoint of phylogeny. – Trans. Amer. ent. Soc., Philadelphia, 52: 199–248.

***Crass, R.S*** (1949): Records of Alderflies (Megaloptera) from Natal and the Eastern Cape, with a description of the larva of *Leptosialis africana* (E.-P.). – Ann. Natal Mus., Pietermaritzburg, 11: 519–522.

***Cummins, K.W.*** (1973): Trophic relations of aquatic insects. – Ann. Rev. Entomol., Palo Alto, Calif., 18: 183–206.

***Cuyler, R.D.*** (1958): The larvae of *Chauliodes* Latreille (Megaloptera: Corydalidae). – Ann. ent. Soc. Amer., Lanham, Maryland, 51: 582–586.

– (1965): The larva of *Nigronia fasciatus* Walker (Megaloptera: Corydalidae). – Ent. News, Philadelphia, 26: 192–194.

***Czihak, G.*** (1953): Beiträge zur Anatomie des Thorax von *Sialis flavilatera* L. – Öst. Zool. Zeits., Wien, 4: 421–448.

***Davis, K.C.*** (1903): Sialididae of North and South America. – Bull. N.Y. State Mus., 68: 442–486.

***Denis, J.R. & Bitsch, J.*** (1973): Morphologie de la tête des insectes, pp. 1–593 in ***Grassé, P.P.*** – (ed.) Traité de Zoologie, Masson, Paris, 8 (1) Paris.

***DuBois, A.M.*** (1936): Recherches experimentales sur la determination de l'embryon dans l'oeuf de *Sialis lutaria*. – Rev. suisse Zool., Genève, 43: 519–523.

– (1938): La determination de l'ebauche embryonnaire chez *Sialis lutaria* L. (Megaloptera). – Rev. suisse Zool., Genève, 45: 1–92.

***DuBois, A.M. & Geigy, R.*** (1935): Beiträge zur Oekologie, Fortpflanzungsbiologie und Metamorphose von *Sialis lutaria* L. (Studien am Sempachersee). – Rev. suisse Zool., Genève 42: 169–248.

***DuFour, L.*** (1841): Recherches anatomiques et physiologiques sur les Orthoptères, les Hymenoptères et les Neuroptères. – Paris.

***Eglin, W.*** (1940): Die Neuropteren der Umgebung von Basel. – Rev. suisse Zool., Genève, 47: 243–358.

***Ehnbom, K.*** (1948): Studies on the central and sympathetic nervous system and some sense organs in the head of Neuropteroid insects. – Opusc. Entomol., Lund, Suppl. 8: 1–162.

***Elliott, J.M.*** (1977): A key to the larvae and adults of British freshwater Megaloptera and Neuroptera with notes on their life cycles and ecology. – Sci. Publns Freshw. Biol. Assoc., Ambleside, No 35.

***Epperson, C.R. & Short, R.A.*** (1987): Annual production of *Corydalus cornutus* (Megaloptera) in the Guadeloupe River, Texas. – Amer. Midl. Nat., Notre Dame, Ind., 118: 433–438.

***Esben-Petersen, P.*** (1924): South African Megaloptera. – Ann. Sth. Afr. Mus., Cape Town 19: 151–158.

***Evans, E.D.*** (1972): A study of the Megaloptera of the Pacific coastal region of the United States. – Ph.D. Thesis, Oregon State University, Corvallis.

***Evans, E.D.*** (1978): Megaloptera and aquatic Neuroptera. Chapter 13 in ***Merritt, R.W. & Cummins, K.W.*** – (eds) An Introduction to the Aquatic Insects of North America. Kendall/Hunt, Dubuque, Iowa.

***Fitch, G.K. & Kammer, A.E.*** (1982): Modulation of the ventilatory rhythm of the hellgrammite *Corydalus cornutus* by mechanosensory input. – J. Comp. Physiol., Heidelberg, 149: 423–434.

***Flint, O.S., Jr.*** (1965): The genus *Neohermes* (Megaloptera: Corydalidae). – Psyche, Cambridge, Mass., 72: 255–263.

– (1970): The Megaloptera of Dominica. – Proc. ent. Soc. Wash., 71: 240–242.

– (1973): The Megaloptera of Chile (Neuroptera). – Revta Chil. Ent., Santiago, 7: 31–45.

– (1983): *Nothochauliodes penae*, a new genus and species of Megaloptera from Chile. (Neuroptera: Corydalidae). – Ent. News, Philadelphia, 94: 15–17.

***Friedrich, H.*** (1953): Neuroptera. In ***Bronns, H.-G.*** – (ed.) Klassen und Ordnungen des Tierreichs. Insecta 12, 148 pp. Leipzig.

***Fuller, R.L. & De Staffan, P.A.*** (1988): A laboratory study of the vulnerability of prey predation by three aquatic insects. – Can. J. Zool., Ottawa, 66: 875–878.

***Fuller, R.L. & Hynes, H.B.N.*** (1987): Feeding ecology of three predacious aquatic insects and two fish in a riffle of the Speed River, Ontario. – Hydrobiologie, Den Haag, 150: 243–255.

***Fuller, R.L. & Rand, P.S.*** (1990): Influence of substrate type on vulnerability of prey to predacious aquatic insects. – J. N. Am. Benthol. Soc., Lawrence, Kansas, 9: 1–8.

***Furnish, J., Belluck, D., Baker, D. & Pennington, B.A.*** (1981): Phoretic relationships between *Corydalus cornutus* (Megaloptera: Corydalidae) and Chironomidae in Eastern Tennessee. – Ann. ent. Soc. Amer. Lanham, Maryland, 74: 29–30.

***Furth, D.G. & Suzuki, K.*** (1990): Comparative morphology of the tibial flexor and extensor tendons in insects. – Syst. Entomol., Oxford, 15: 433–441.

***Gams, H.*** (1927): Die Geschichte der Lunzer Seen, Moore und Wälder. – Int. Rev. Hydrobiol., Leipzig, 18: 305–387.

***Geigy, R.*** (1937): Beobachtungen über die Metamorphose von *Sialis lutaria* L. – Jahres. S. E. G., Basel 1937: 144–157.

***Geigy, R. & DuBois, A.M.*** (1935): Sinnesphysiologische Beobachtungen über die Begattung von *Sialis lutaria* L. – Rev. suisse Zool., Genève, 42: 447–457.

***Geigy, R. & Grobe, D.*** (1958): Die ökologische Abhängigkeit des Metamorphose-Geschehens bei *Sialis lutaria* L. – Rev. suisse Zool., Genève, 43: 519–523.

***Geigy, R. & Ochse, W.*** (1940a): Schnürungsversuche an Larven von *Sialis lutaria* L. – Rev. suisse Zool., Genève, 47: 193–194.

– (1940b): Versuche über die inneren Faktoren der Verpuppung von *Sialis lutaria* L. – Bull. Biol., Dorpat, 47: 225–241.

***Geigy, R. & Rahm, U.H.*** (1951): Beiträge zur experimentellen Analyse der Metamorphose von *Sialis lutaria* L. – Bull. biol., Dorpat, 58: 408–413.

***Geijskes, D.C.*** (1986). Notes on Megaloptera from the Guyanas, S. Am. In ***Gepp, J., Aspöck, H. & Hölzel, H.*** – (eds) Progress in World's Neuropterology, pp. 79–84. Graz.

***Giani, N. & Laville, H.*** (1973): Cycle biologique et production de *Sialis lutaria* L. (Megaloptera) dans le Lac de Port-Bielh (Pyrenees centrales). – Ann. Limnol., Paris, 9: 45–61.

***Glorioso, M.J.*** (1981): Systematics of the dobsonfly subfamily Corydalinae (Megaloptera: Corydalidae). – Syst. Entomol., Oxford, 6: 253–290.

***Glorioso, M.J. & Flint, O.S.***, Jr. (1984): A review of the genus *Platyneuromus* (Insecta: Neuroptera: Corydalidae). – Proc. Biol. Soc., Wash., 97: 601–614.

***Gorayeb, I.S. & Pinger, R.R.*** (1978): Deteccao de predatores naturais das larvas de *Simulium fulvinotum* Cerq. e Mello, 1968 (Diptera, Nematocera). – Acta Amazonica, Manaus, 8: 629–637.

***Gotceitas, V. & Mackay, R.J.*** (1980): The phoretic association of *Nanocladius (Nanocladius) rectinervis* (Kieffer) (Diptera: Chironomidae) on *Nigronia serricornis* Say (Megaloptera: Corydalidae). – Can. J. Zool., Ottawa, 58: 2260–2263.

***Gravenhorst, J.L.C.*** (1834): Bericht über die in Bernstein erhaltenen Insekten der phys.-ökon. Gesellschaft in Königsberg. – Arb. schles. Gesellsch. f. vaterl. Kultur, Breslau 1834: 92–93.

***Grenacher, H.*** (1879): Untersuchungen über das Sehorgan der Arthropoden. – Göttingen.

***Gross, J.*** (1903): Untersuchungen über die Histologie des Insektenovariums. – Zool. Jahrb. Anat., Jena, 18: 71–186.

***Hagen, H.*** (1854): Über die Neuropteren der Bernsteinfauna. – Verh. Zool. Bot. Ges. Wien 4: 221–232.

– (1856): Die im Bernstein befindlichen Neuropteren der Vorwelt. – Nicholaische Buchhandlung, Berlin.

***Haldeman, S.S.*** (1848): History and transformation of *Corydalus cornutus*. – Mem. Amer. Acad. Arts Sci., Cambridge, 4: 157–161.

***Hammar, A.G.*** (1908): On the nervous system of the larva of *Corydalis cornuta* L. – Ann. ent. Soc. Amer., Lanham, Maryland, 1: 105–127.

***Hamilton, A.*** (1940): The New Zealand Dobson-fly (*Archichauliodes diversus* Walk.): life-history and bionomics. – N.Z.J. Sci. Technol., Wellington, 22: 44A–55A.

***Handlirsch, A.*** (1908): Die fossilen Insekten und die Phylogenie der rezenten Formen. – Engelman, Leipzig.

***Handlirsch, A. & Beier, M.*** (1936): Megaloptera = Schlammfliegen. In: ***Kukenthal, W. & Krumbach, T.*** – Handbuch der Zoologie, 4: 1384–1401.

***Hanstrom, B.*** (1940): Inkretorische Organe, Sinnesorgane und Nervensystem des Kopfes einiger niederer Insektenordnungen. – K. svensk. Vertensk.-Akad. Handl. Stockholm 18, 266 p.

***Hassan, A.A.G.*** (1944): The structure and mechanism of the spiracular regulatory apparatus in adult Diptera and certain other groups of insects. – Trans. Roy. ent. Soc. Lond., 94: 103–154.

***Hayashi, F.*** (1985): Factors affecting the handling time of the dobsonfly larva, *Protohermes grandis* (Megaloptera, Corydalidae). – Jap. J. Ecol. Senohai, 35: 1–11.

– (1988a): Lifehistory variation in a dobsonfly, *Protohermes grandis* (Megaloptera: Corydalidae: effects of prey availability and temperature. – Freshw. Biol., Oxford, 19: 205–216.

– (1988b): Prey selection by the dobsonfly larva, *Protohermes grandis* (Megaloptera: Corydalidae). – Freshw. Biol., Oxford, 20: 19–29.

– (1989): Microhabitat selection by the fishfly larva, *Parachauliodes japonicus*, in relation to its mode of respiration. – Freshw. Biol., Oxford, 21: 489–496.

– (1990a): Convergence of insular dwarfism in damselflies (*Euphaea*) and dobsonflies (*Protohermes*). – Freshw. Biol., Oxford, 23: 219–231.

– (1990b): Notes on *Neochauliodes sinensis* (Walker) (Megaloptera, Corydalidae) on Amami-Oshima Island, Kagoshima Prefecture, Japan. – Gekkan-Mushi, Tokyo, (228): 14–15.

– (1990c): Life history and distribution of *Parachauliodes japonicus* (McLachlan). – Collecting & Breeding, Tokyo, 52: 396–399.

***Hayashi, F. & Nakane, M.*** (1988): A radio-tracking study on the foraging movements of the dobsonfly larva, *Protohermes grandis* (Megaloptera, Corydalidae). – Kontyu, Tokyo, 56: 417–429.

– (1989): Radio tracking and activity monitoring of the dobsonfly larva, *Protohermes grandis* (Megaloptera: Corydalidae). – Oecologia, Berlin, 78: 468–472.

***Hayashi, F. & Yoshida, T.*** (1987): Seasonal change in the respiration of the dobsonfly larva, *Protohermes grandis* (Megaloptera: Corydalidae). – Freshw. Biol., Oxford, 18: 117–124.

***Hennig, W.*** (1981): Insect Phylogeny. – John Wiley, Chichester, 514 pp.

***Hesse, R.*** (1901): Untersuchungen über Organe der Lichtempfindung bei niederen Tieren. VII. Von den Arthropodenaugen. – Z. wiss. Zool., Leipzig, 70: 347–473.

***Hildrew, A.G. & Townsend, C.R.*** (1982): Predators and prey in a patchy environment: a freshwater study. – J. anim. Ecol., Oxford, 51: 797–815.

***Hilsenhoff, W.L.*** (1968): Phoresy by *Plecopteracoluthus downesi* on larvae of *Nigronia serricornis*. – Ann. ent. Soc. Amer., Lanham, Maryland, 61: 1622–1623.

***Hilton, W.A.*** (1909): The tracheal study in the central nervous system of the larva of *Corydalis cornuta* L. – Ann. ent. Soc. Amer., Lanham, Maryland, 2: 46–59.

– (1911): The structure of the central nervous system of *Corydalis* larva. – Ann. ent. Soc. Amer., Lanham, Maryland, 4: 219–260.

***Hinton, H.E*** (1981): Biology of Insect Eggs. 3 vols. – Pergamon Press, Oxford.

***Hitchcock, E.*** (1858): Ichnology of New England. A report on the sandstone of the Connecticuit Valley, especially its fossil footmarks. – Boston. 1–220.

***Howard, L.O.*** (1896): A coleopterous enemy of *Corydalis cornutus*. – Proc. ent. Soc. Wash. 3: 310–313.

***Hughes-Schrader, S.*** (1980): Segregational mechanisms of sex chromosomes in Megaloptera (Neuropteroidea). – Chromosoma, Berlin, 81: 307–314.

***Illies, J.*** (1967): Megaloptera and Plecoptera (Ins.) aus den jungpliozänen Süsswassermergeln von Willershausen. – Ber. Naturhist. Ges., Nürnberg, 111: 47–55.

***Itoh, H.*** (1933a): The chromosomes of *Chauliodes japonica* MacLachlan. – Zool. Mag., Tokyo, 45: 1–12.

– (1933b): The chromosomes of *Protohermes grandis* Thunberg. – Zool. Mag., Tokyo, 45: 199–210.

***Iversen, T.M. & Thorup, J.*** (1987): Population dynamics and production of *Sialis lutaria* L. (Megaloptera) in the Danish River Susa. – Freshw. Biol., Oxford, 17: 461–469.

***Iversen, T.M., Thorup, J., Hansen, T., Lodal, J. & Olsen, J.*** (1985): Quantitative estimates and community structure of invertebrates in a macrophyte rich stream. – Arch. Hydrobiol., Stuttgart, 102: 291–301.

***Jander, U.*** (1966): Untersuchungen zur Stammesgeschichte der Putzbewegungen von Tracheaten. – Z. Tierpsychol., Berlin, 23: 799–844.

***Jones, R.C. & Clark, C.C.*** (1987): Impact of watershed urbanization on stream insect communities. – Water Resources Bull., Salt Lake City, Utah, 23: 1047–1055.

***Kaiser, E.W.*** (1961): Studier over de danske *Sialis*-arter. II (Megaloptera). Biologien hos *S. fuliginosa* Pict. og *S. nigripes* Ed. Pict. – Flora Fauna, Aarhus, 67: 74–96.

***Kelsey, L.P.*** (1954): The skeleto-motor mechanism of the Dobson fly, *Corydalus cornutus* Part I. Head and prothorax. – Cornell Univ. Agr. Exp. Sta. Mem., Geneva, 334: 1–52.

– (1957): The skeleto-motor mechanism of the Dobson Fly, *Corydalus cornutus* Part II. Pterothorax. – Cornell Univ. Agr. Exp. Sta. Mem., Geneva, 346: 1–42.

***Kevan, D.K. McE*** (1979): 37. Megaloptera pp 351–352. In ***Danks, H.V.*** (ed.) Canada and its Insect Fauna. – Mem. ent. Soc. Can., Ottawa, 108.

***Khalifa, A.*** (1949): Spermatophore production in Trichoptera and some other insects. – Trans. Roy. ent. Soc. Lond., 100: 449–471.

***Killington, F.J.*** (1932a): A contribution towards a biological survey of the Megaloptera and Neuroptera of Hampshire and the Isle of Wight. – Trans. ent. Soc. Sth. Engl., Southampton, 7: 57–67.

– (1932b): On the pairing of *Sialis fuliginosa* Pict. (Neuroptera: Megaloptera). – Entomologist, London, 65: 66–67.

***Kimmins, D.E.*** (1948): Notes on the genus *Protohermes* Weele (Megaloptera), with descriptions of two new species. – Ann. Mag. nat. Hist., London, 12 (1). 765–781.

– (1954): A new genus and some new species of the Chauliodini (Megaloptera) with notes on certain previously described species. – Bull. Brit. Mus. (Nat. Hist.), Entomol., London, 3: 417–444.

***King, R.C. & Büning, J.*** (1985): The origin and functioning of insect oocytes and nurse cells. In ***Kerkut, G.A. Gilbert, L.I.*** Comprehensive Insect Physiology Biochemistry and Pharmacology. Vol. 1. Embryogenesis and Reproduction. pp. 37–82. – Pergamon Press, Oxford.

***Kinnamon, S.C. & Kammer, A.E.*** (1983): Neural control of ventilatory movements in the aquatic insect *Corydalus cornutus*: the motor pattern. - J. Comp. Physiol., Heidelberg, A 153: 543–555.

***Kinnamon, S.C., Kammer, A.E. & Kiorpes, A.L.*** (1980): Control of ventilation in an aquatic insect larva: central effect of hypoxia. – Amer. Zool., Bloomington, Ind., 20: 942.

– (1984): Control of ventilatory movements in the aquatic insect *Corydalus cornutus:* central effect of hypoxia. – Physiol. Entomol., Oxford, 9: 19–28.

***Klingstedt, H.*** (1937): Chromosome behaviour and phylogeny in the Neuroptera. – Nature, London, 139: 468–469.

***Knight, A.W. & Simmons, M.A.*** (1975a): Factors influencing the oxygen consumption of the hellgrammite, *Corydalus cornutus* (L.) (Megaloptera: Corydalidae). – Comp. Biochem. Physiol., Oxford, 50A: 827–833.

– (1975b): Factors influencing the oxygen consumption of larval *Nigronia serricornis* (Say) (Megaloptera: Corydalidae). – Comp. Biochem. Physiol., Oxford, 51A: 117–123.

***Kramer, S.*** (1955): The musculature of the head of the *Corydalus* larva (Neuroptera, Sialidae). – J. Morphol., Philadelphia, 96: 1–30.

***Krauss, W.C.*** (1884): On the nervous system of the head of the larva of *Corydalus cornutus* Linn. – Psyche, Cambridge, Mass., 4: 179–184 (and plate 2).

***Kristensen, N.P.*** (1981): Phylogeny of insect orders. – Ann. Rev. Ent., Palo Alto, Calif., 26: 135–157.

***Kuwayama, S.*** (1962): A revisional synopsis of the Neuroptera in Japan. – Pacif. Insects., Honolulu, 4: 325–412.

– (1964): On the Neuroptera from Amami – Oshima and Yakushima. – Mushi, Fukuoka, 38: 25–32.

– (1965): The genus *Sialis* from Shikoku and Kyushu, Japan (Neuroptera: Sialidae). – Trans. Shikoku ent. Soc., Matsuyama, 8: 87–90.

***Larsson, Sv.G.*** (1978): Baltic amber, a palaeobiological study. – Entomonograph, Klampenborg, 1: 1–92.

***Lassiere, O.L.*** (1988): Host-parasite relationships between larval *Sialis lutaria* (Megaloptera) and *Neoechinorhynchus rutili* (Acanthocephala). – Parasitology, Cambridge, 97: 331–338.

***Latreille, P.A.*** (1796): Précis des caractères géneriques des Insectes, disposes dans un ordre naturel par le Citoyen Latreille. – Brive, Bordeaux.

– (1802): Histoire naturelle, génerale et particulière, des Crustaces et des Insectes. 3. – Dufart, Paris. 467 pp.

– (1803): Histoire naturelle, génerale et particulière, des Crustaces et des Insectes. 5. – Dufart, Paris, 406 pp.

– (1804): Histoire naturelle, génerale et particulière, des Crustaces et des Insectes. 9. – Dufart, Paris.

***Lattin, J.D.*** (1952): (unpublished observations cited by ***Evans, E.D.*** 1972).

***Leidy, J.*** (1848): Internal anatomy of *Corydalus cornutus* in its three stages of existence. – Mem. Am. Acad. Arts Sci., Cambridge, Mass., 4: 162–168.

***Leischner, T.G. & Pritchard, G.*** (1973): The immature stages of the alderfly, *Sialis cornuta* (Megaloptera: Sialidae). – Can. Ent., Ottawa, 105: 411–418.

***Lestage, J.A.*** (1919): Notes biologiques sur *Sialis lutaria* L. (Megaloptera). – Ann. Biol. lacustre, Bruxelles, 9: 25–38.

– (1927): La faune entomologique indo-chinoise, 2: les Megaloptères. – Ann. Soc. ent. Belg., Bruxelles, 67: 71–90, 93–119.

***Lilly, C.K., Ashley, D.L. & Tarter, D.C.*** (1978): Observations on a population of *Sialis itasca* Ross in West Virginia (Megaloptera: Sialidae). – Psyche, Cambridge, Mass., 85: 209–217.

***Linnaeus, C.*** (1758): Systema Naturae per Regna Tria Naturae Secundam Classes, Ordines, Genera, Species, cum Characteribus, differentiis, Synonymis, Locis. – 10th ed. Holmiae.

– (1763): Centuria Insectorum rariorum. – Amoen Acad., 6: 384–415. Upsala.

***Lintner, J.A.*** (1892): *Chauliodes pectinicornis, rastricornis* and *serricornis*. – 8th Ann. Rpt. N.Y. State Entomologist: 155–159.

***Loew, H.*** (1848): Abbildungen und Bemerkungen zur Anatomie einiger Neuropterengattungen. – Linn. Ent., Berlin, 3: 345–386.

***Maddux, D.C.*** (1952): A study of the dobsonflies (Megaloptera) of the Chico, California area. – M.S. Thesis, Chico State College, Chico, California.

***Maki, T.*** (1936): Studies on the skeletal structure and nervous system of the alderfly *Chauliodes formosanus* Petersen. – Mem. Fac. Sci. Agric., Taihoku Imp. Univ. Formosa, 16: 117–243.

***Mansell, M.W. & Aspöck, H.*** (1990): Preface/Vorwort/Avan-Propos/Voorwoord. In ***Mansell, M.W. & Aspöck, H.*** (eds.). – Advances in Neuropterology, pp. 1–6, Department of Agricultural Development, Pretoria.

***Martynov, A.*** (1928): Permian fossil insects of North-East Europe. – Trav. Mus. geol. USSR Acad. Sci., Leningrad, 4: 1–118.

***Martynova, O.M.*** (1952): Permian lacewings from the USSR. – Trudy palaeont. Inst., Leningrad, 40: 187–237.

– (1961): Order Miomoptera. – Trudy palaeont. Inst., Leningrad, 85: 234–244.

***Mather, T.N.*** (1981): Larvae of alderfly (Megaloptera: Sialidae) from pitcher plant. – Ent. News, Philadelphia, 92: 32.

***Matsuda, R.*** (1965): Morphology and evolution of the insect head. – Mem. Amer. ent. Inst. 4. Ann Arbor, Michigan, 334 pp.

– (1970): Morphology and evolution of the insect thorax. – Mem. ent. Soc. Can., Ottawa, 76.

– (1976): Morphology and evolution of the insect abdomen. – Pergamon Press, Oxford.

***Matsuzaki, M. & Ando, H.*** (1977): Ovarian structures of the adult alderfly, *Sialis mitsuhashii* Okamoto (Megaloptera: Sialidae). – Int. J. Ins. Morphol. Embryol., Oxford, 6: 17–29.

***Matsuzaki, M., Sasaki, A. & Komuro, M.*** (1985): Panoistic ovarioles of the dobsonfly *Protohermes grandis* (Megaloptera, Corydalidae). In ***Ando, H. & Miya, K.*** (eds). – Recent Advances in Insect Embryology in Japan, pp. 13–23. ISEBU, Tsukuba.

***McLachlan, R.*** (1869): Considerations on the neuropterous genus *Chauliodes* and its allies; with notes and descriptions. – Ann. Mag. nat. Hist., London, 4 (4): 35–46.

– (1899) A second Asiatic species of *Corydalis*. – Trans. ent. Soc. Lond., 1900: 281–283.

***Michael, D.L. & Culver, D.A.*** (1987): Influence of plecopteran and megalopteran predators on *Hydropsyche* (Trichoptera: Hydropsychidae) microdistribution and behaviour. – J.N. Am. Benthol. Soc., Lawrence, Kansas, 6: 46–55.

***Mickoleit, G.*** (1973): Über den Ovipositor der Neuropteroidea and Coleoptera und seine phylogenetische Bedeutung (Insecta: Holometabola). – Zeits. Morphol. Tiere, Berlin, 74: 37–64.

***Miyakawa, K.*** (1979): Embryology of the dobsonfly, *Protohermes grandis* Thunberg (Megaloptera: Corydalidae). I. Changes in external form of the embryo during development. – Kontyu, Tokyo, 47: 367–375.

– (1980): Embryogenesis of the pleuropodia and the abdominal filaments (tracheal gills) in *Protohermes grandis* Thunberg (Megaloptera: Corydalidae). – XVIth Int. Congr. Entomol., Abstracts, p. 50.

***Monserrat, V.J.*** (1986): Longinos Navás, his neuropterological work and collection. In ***Gepp, J., Aspöck, H. & Hölzel, H.*** (eds). – Recent Research in Neuropterology, pp 173–176. Graz.

***Moody, H.L.*** (1877/78): The larva of *Chauliodes*. – Psyche, Cambridge, Mass., 2: 52–53. (issued Jan 12 1878, but volume dated 1877).

***Navás, L.*** (1925): Neuropteren, Megalopteren, Plecopteren und Trichopteren aus dem Deutschen Entomologischen Institut (Berlin-Dahlem). – Ent. Mitt., Berlin, 14: 205–212.

– (1927): Veinticinco formas nuevas de insectos. – Bol. Soc. Iber. Cienc. Nat., Zaragoza, 26: 48–75.

***Needham, J.G. & Betten, C.*** (1901): Aquatic insects of the Adirondacks. – Bull. N.Y. State Mus. 47: 383–612.

***Nel, A.*** (1988): Les Sialidae (Megaloptera) fossiles des diatomites de Murat (Cantal, France) et de Bes-Konak (Anatolie, Turque). – Neur. Int., Nice, 5: 39–44.

***Neunzig, H.H.*** (1966): Larvae of the genus *Nigronia* Banks (Neuroptera: Corydalidae). – Proc. ent. Soc. Wash., 68: 11–16.

***New, T.R.*** (1989): Planipennia (Lacewings). – Handbuch der Zoologie, 4 (30): 1–132.

***Nielsen, L.A. & Orth, D.J.*** (1988): The hellgrammite-crayfish bait fishery of the New River and its tributaries, West Virginia. – N. Amer. J. Fish. Management, Bethesda, Md. 8: 317–324.

***Ochse, W.*** (1944): Experimentelle und Histologische Beiträge zur inneren Metamorphose von *Sialis lutaria* L. – Rev. suisse Zool., Genève, 51: 1–82.

***Okamoto, N.*** (1910): Die Sialiden Japans. – Wien. Ent. Ztg., 29: 255–263.

***Panov, A.A. & Davydova, E.D.*** (1976): Medial neurosecretory cells in the brain of Mecoptera and Neuropteroidea (Insecta). – Zool. Anz., Leipzig, 197: 187–206.

***Parfin, S.*** (1952): Notes on the bionomics of *Corydalus cornutus* (Linne), *Chauliodes rastricornis* Rambur, *C. pectinicornis* (Linne) and *Neohermes* sp.. – Amer. Midl. Nat., Notre Dame, Ind., 47: 426–434.

***Paul, H. & Rouff, S.*** (1927): Pollenstatistische und stratigraphische Mooruntersuchungen im südlichen Bayern. I. Teil.-Ber. bayer. bot. Ges., München, 19: 1–84.

– (1932): Pollenstatistische und stratigraphische Mooruntersuchungen im südlichen Bayern. II Teil. – Ber. bayer. bot. Ges., München, 20: 1–264.

***Paulian, R.*** (1951): Faune des eaux douces de Madagascar. Plecoptères et Megaloptères. – Mem. Inst. Sci. Madagascar Ser A., 6: 53–61.

***Paulus, H.F.*** (1986): Comparative Morphology of the larval eyes of Neuropteroidea. In: ***Gepp, J., Aspöck, H. & Hölzel, H.*** (eds). – Recent Research in Neuropterology, pp. 157–164, Graz.

***Penland, D.R.*** (1953): A detailed study of the life cycle and respiratory system of a new species of western dobsonfly, *Neohermes aridus*. – M.S. Thesis, Chico State College, Chico, California.

***Penny, N.D.*** (1977): Lista de Megaloptera, Neuroptera e Raphidioptera do Mexico, America Central, ilhas Caraibas e America do Sul. – Acta Amazonica, Manaus, 7, Supplement.

– (1981): Neuroptera of the Amazon Basin. Part 4. Sialidae. – Acta Amazonica, Manaus, 11: 843–846.

– (1982): Neuroptera of the Amazon Basin. Part 7. Corydalidae. – Acta Amazonica, Manaus, 12: 825–837.

***Penny, N.D. & Flint, O.S.***, Jr. (1982): A revision of the genus *Chloronia* (Neuroptera: Corydalidae). – Smiths. contrib. Zool., Washington, No. 348: 1–27.

***Peterson, A.*** (1960): Larvae of Insects. An introduction to Nearctic species. Part II. – Columbus, Ohio.

***Phillips, J. S.*** (1929): A report on the food of trout and other conditions affecting their wellbeing in the Wellington district. – Fisheries Bulletin No. 2. Wellington, New Zealand.

***Pictet, F. J.*** (1836): Memoire sur le genre *Sialis* de Latreille, et considerations sur le classification de l'ordre des Neuroptères. – Ann. Sci. nat., Paris, 5: 69–80.

– (1856): Verbreitung der Neuroptera im Bernstein. In: ***Berendt, G. C.*** Die im Bernstein befindlichen organischen Reste der Vorwelt gesammelt in Verbindung mit Mehreren bearbeitet und herausgegeben von ***G. C. Behrendt***. – II. Nicolai, Berlin.

***Ping, C.*** (1928): Cretaceous fossil insects of China. – Palaeont. sinica. ser. B., 13: 1–56.

***Ponomarenko, A. G.*** (1961): The systematic position of *Coptoclava longipoda* Ping (Insecta, Coleoptera). – Palaeont. Zhur., Moscow, 3: 67–72.

– (1976): Corydalidae (Megaloptera) from the Cretaceous of northern Asia. – Ent. Obozr., Moskva, 55: 425–433. (English translation: Ent. Rev. 55: 114–122).

– (1977): Palaeozoic members of Megaloptera (Insecta). – Palaeont. Zh. 1977 (1): 78–86. (English translation: Palaeont. J., Washington, 11: 73–81).

***Pritchard, G. & Leischner, T. G.*** (1973): The life history and feeding habits of *Sialis cornuta* Ross in a series of abandoned beaver ponds (Insecta: Megaloptera). – Can. J. Zool., Ottawa, 51: 121–131.

***Rahm, U. H.*** (1952): Die innersektorische Steuerung der postembryonalen Entwicklung von *Sialis lutaria* L. (Megaloptera). – Rev. suisse. Zool, Genève, 59: 173–237.

***Rambur, P.*** (1842): Histoire naturelle des Insectes Neuroptères. – Paris.

***Rasnitsyn, A. P.*** (1977): New Palaeozoic and Mesozoic insects. – Palaeont. Zhur., Moscow, 1: 74–77.

***Rath, O. vom*** (1888): Über die Hautsinnesorgane der Insekten. – Z. wiss. Zool., Leipzig, 46: 413–454.

***Richards, O. W. & Davies, R. G.*** (1957): Imms' General Textbook of Entomology. 9th (revised) edition. – Methuen, London.

***Riek, E. F.*** (1954): The Australian Megaloptera or Alderflies. – Aust. J. Zool., Melbourne, 2: 131–142.

– (1968): Undescribed fossil insects from the Upper Permian of Belmont, New South Wales. – Rec. Aust. Mus., Sydney, 27: 303–310.

– (1970a): Fossil History. In: CSIRO, The Insects of Australia, pp. 168–186. – Melbourne University Press, Melbourne.

– (1970b): Megaloptera. In: CSIRO, The Insects of Australia, pp. 465–471. – Melbourne University Press, Melbourne.

– (1974): Upper Triassic insects from the Molteno 'Formation', South Africa. – Palaeontol. afr., Johannesburg, 17: 19–31.

– (1975): On the phylogenetic position of *Brucheiser argentinus* Navás 1927 and description of a second species from Chile (Insecta: Neuroptera). – Stud. neotrop. Fauna, Lisse, 10: 117–126.

– (1976): New Upper Permian insects from Natal, South Africa. – Ann. Natal Mus., Pietermaritzburg, 22: 755–789.

***Riley, C. V.*** (1873): The hellgrammite fly – *Corydalus cornutus* (Linn.). In: Fifth anual report on the noxious, beneficial, and other insects of the state of Missouri, pp. 142–145.

– (1876): On the curious egg mass of *Corydalus cornutus* (Linn). – Proc. Amer. Assoc. Adv. Sci., Philadelphia, 25: 275–279.

– (1877): The hellgrammite – *Corydalus cornutus* (Linn.) In: Ninth annual report on the noxious, beneficial, and other insects of the state of Missouri, pp. 125–129.

– (1878): On the larval characteristics of *Corydalis* and *Chauliodes* and on the development of *Corydalis cornutus*. – Proc. Amer. Assoc. Adv. Sci., Philadelphia, 27: 285–287.

– (1879): (same title as 1878 paper, but Can. Ent., Ottawa, 11: 96–98).

***Roback, S. S.*** (1974): Insects (Arthropoda: Insecta). In ***Hart, C. W. Jr. & Fuller, S. L. H.*** (eds) Pollution ecology of freshwater invertebrates pp 313–386. – Academic Press, New York and London.

***Röber, H.*** (1941): Morphologie des Kopfes und des Vorderdarmes der Larve und Imago von *Sialis flavilatera*. – Zool. Jahrb. Anat., Jena, 67: 61–118.

***Rohdendorf, B. B.*** (1969): Insect phylogeny and palaeontological data. – Ent. Obozr., Moskva, 47: 322–342.

***Ross, H. H.*** (1937): Studies of Nearctic aquatic insects. I. Nearctic Alder Flies of the genus *Sialis* (Megaloptera, Sialidae). – Bull. Ill. nat. Hist. Surv., Illinois, 21: 57–78.

***Rupprecht, R.*** (1975): Die Kommunikation von *Sialis* (Megaloptera) durch Vibrationssignale. – J. Ins. Physiol., Exeter, 21: 305–320.

***Salt, G.*** (1937): The egg parasite of *Sialis lutaria*: a study of the influence of the host upon a dimorphic parasite. – Parasitology, Cambridge, 29: 539–553.

***Saunders, W.*** (1875): The Hellgrammite fly – *Corydalis cornutus* Linn. – Can. Ent., Ottawa, 7: 64–66.

***Scudder, S. H.*** (1878): An account of some insects of unusual interest from the Tertiary rocks of Colorado and Wyoming. – Bull. U. S. Geol. Surv., Reston, Virg., u. s. w., 4: 519–541.

***Seitz, W.*** (1940): Zur Frage des Extremitätencharakters der Tracheenkiemen von *Sialis flavilatera* L. im Rahmen allgemeiner biologischer Untersuchungen. – Zeits. Morph. Ökol. Tiere, Berlin, 37: 214–275.

***Selman, B. J.*** (1960a): On the tissue isolated in some of the larval appendages of *Sialis lutaria* L. at the larval-pupal moult. – J. Ins. Physiol., Exeter, 4: 235–257.

– (1960b): Tolerance of dehydration of the blood of *Sialis lutaria* L. – J. Ins. Physiol., Exeter, 6: 81–83.

– (1962): The fate of the blood cells during the life history of *Sialis lutaria* L. – J. Ins. Physiol., Exeter, 8: 209–214.

– (1965): The circulatory system of the alder fly *Sialis lutaria*. – Proc. Zool. Soc. Lond., 144: 487–535.

***Sharov, A. G.*** (1953): First discovery of a Permian larva of a sialid-fly (Megaloptera) from Kargala. – Dokl. Akad. Nauk SSSR., Petrograd, 89: 731–732.

***Shaw, J.*** (1955a): The permeability and structure of the cuticle of the aquatic larva of *Sialis lutaria*. – J. exp. Biol., Cambridge, 32: 330–352.

– (1955b): Ionic regulation and water balance in the aquatic larva of *Sialis lutaria*. – J. exp. Biol., Cambridge, 32: 353–382.

***Short, R. A., Stanley, E. H., Harrison, J. W. & Epperson, C. R.*** (1987): Production of *Corydalus cornutus* (Megaloptera) in four streams differing in size, flow, and temperature. – J. N. Am. Benthol. Soc., Lawrence, Kansas, 6: 105–114.

***Silvestri, F.*** (1908): Sviluppo dell' *Oophthora semblidis* Aur. – Boll. Lab. zool. Portici, 3: 72–83.

***Smith, E. L.*** (1970): Biology and structure of the dobsonfly, *Neohermes californicus* (Walker) (Megaloptera: Corydalidae). – Pan.-Pac. Ent., San Francisco, 46: 142–150.

***Smith, R.C.*** (1920): The process of hatching in *Corydalis cornuta* Linn. – Ann. ent. Soc. Amer., Lanham, Maryland, 13: 70–74.

– (1922): Hatching in three species of Neuroptera. – Ann. ent. Soc. Amer., Columbus, 15: 169–176.

***Snodgrass, R.E.*** (1935): Principles of insect morphology. – McGraw-Hill, New York.

– (1957): A revised interpretation of the external reproductive organs of male insects. – Smiths. Misc. Coll., Washington, 135: 1–60.

***Staddon, B.W.*** (1955): The excretion and storage of ammonia by the aquatic larva of *Sialis lutaria* (Neuroptera). – J. exp. Biol., Cambridge, 32: 84–94.

***Stewart, K.W., Friday, G.P. & Rhame, R.E.*** (1973): Food habits of hellgrammite larvae, *Corydalus cornutus* (Megaloptera: Corydalidae), in the Brazos River, Texas. – Ann. ent. Soc. Amer., Lanham, Maryland, 66: 959–963.

***Stitz, H.*** (1908): Zur Kenntnis des Genitalapparats der Neuropteren. – Zool. Jahrb., Anat., Jena, 27: 377–448.

– (1926): Megaloptera. In ***Schulze, P.*** – (ed.) Biologie der Tiere Deutschlands, 35: 29–47.

***Stitz, J.*** (1914): Sialiden der Sammlung des Berliner Museums. – Sitz, Gess. Naturf. Freunde zu Berlin 1914: 191–205.

***Strindberg, H.*** (1916): Hauptzüge der Entwicklungsgeschichte von *Sialis lutaria* L. (Eine embryologische Untersuchung). – Zool. Anz., Leipzig, 46: 167–185.

***Suckow*** (no initial) (1828): *Semblis bicaudata* Fabr. und *Semblis lutaria* Fabr. Zwei verschiedene Gattungen (Genera). – Zeits. organ. Physik, Halle, 2: 265–272.

***Suzuki, N., Shimizu, S. & Ando, H.*** (1981): Early embryology of the alderfly, *Sialis mitsuhashii* Okamoto. – Int. J. Ins. Morphol. Embryol., Oxford, 10: 409–418.

– (1987): Embryonic development of nervous system in the alderfly, *Sialis mitsuhashii* Okamoto (Megaloptera, Sialidae). In ***Ando, H. & Jura, Cz.*** (eds). – Recent Advances in Insect Embryology in Japan and Poland, pp225–235. ISEBU, Tsukaba (Arthr. Embryol. Soc. Japan).

***Symes, E.E.*** (1935): Biological notes on British Megaloptera. – Proc. Trans. S. Lond. ent. nat. Hist. Soc. 1934–35: 121–124.

***Takenaka, M.*** (1975): Corydalidae larvae. – Biol. J. Nara. Womens Univ., Nara, 24–25: 13–14.

***Tarter, D.C., Watkins, W.D. & Little, M.L.*** (1975): Life history of the fishfly, *Nigronia fasciatus* (Megaloptera: Corydalidae). – Psyche, Cambridge, Mass., 82: 81–88.

***Tarter, D.C., Watkins, W.D., Little, M.L. & Ashley, D.L.*** (1977): Seasonal emergence patterns of fishflies east of the Rocky Mountains (Megaloptera: Corydalidae). – Ent. News, Philadelphia, 88: 69–76.

***Tarter, D.C. & Woodrum, J.E.*** (1972): Low pH tolerance of the larvae of the Alderfly, *Sialis aequalis* Banks, under controlled conditions. – Proc. W. Va Acad. Sci., Morgantown, 33: 85–88.

***Theischinger, G.*** (1983): The adults of the Australian Megaloptera. – Aquat. Ins., Lisse, 5: 77–98.

– (1991): Megaloptera. In CSIRO, The Insects of Australia. 2nd ed. I: 516–520 – Melbourne University Press, Melbourne.

***Theischinger, G. & Houston, W.W.K.*** (1988): Megaoptera. In ***Walton, D.W.*** – (ed.) Zoological Catalogue of Australia, 6: 23–32. Australian Government Publishing Service, Canberra.

***Thorpe, W.H.*** (1933): Experiments on the respiration of aquatic and parasitic insect larvae. – Proc. Vth Int. Congr. Entomol., Paris, 2: 345–351.

***Thunberg, C.P.*** (1781): Dissertatio Entomologica novas Insectorum species sistens. – Edman, Upsala.

***Tillyard, R.J.*** (1918): Australian Megaloptera or alderflies, descriptions of new genera and species. – Proc. Linn. Soc. N.S.W., Sydney, 43: 819–825.

***Tjeder, B.*** (1944): A note on the food of the adult *Sisyra fuscata* F. (Neuroptera: Sisyridae). – Ent. Tidskr., Stockholm, 65: 203–204.

– (1968): A subfossil find of *Sialis* in Sweden. – Opusc. Entomol., Lund, 33: 189–190.

– (1970): Neuroptera. In: ***Tuxen, S.L.*** – (ed.) Taxonomists Glossary of Genitalia in Insects. 2nd ed. Copenhagen, pp. 89–99.

– (1974): Class Insecta, order Neuroptera. In Status of the taxonomy of the Hexapoda of southern Africa. – Ent. Mem., Dept. Agric. S. Afr., Pretoria, 38: 77–79.

***Tracy, B.H. & Hazelwood, D.H.*** (1983): The phoretic association of *Urnatella gracilis* (Entoprocta: Urnatellidae) and *Nanocladius downesi* (Diptera: Chironomidae) on *Corydalus cornutus* (Megaloptera: Corydalidae). – Freshw. Invert. Biol., Milwaukee, WI., 2: 186–191.

***Van der Weele, H.W.*** (1906): Übersicht der Sialiden des Indo-Malayischen Archipels. – Notes Leyd. Mus., Leiden, 26: 207–222.

– (1910): Megaloptera. Cat. Coll. Zool. Sel.-Long. Brussels, 5.

***Villot, A.*** (1885): Sur l'état larvaire et l'hôte intermediaire de l'*Echinococcus clavaeceps* Zeder. – Zool Anz., Leipzig, 8: 19–22.

***Voss, R.S.*** (1988): Systematics and ecology of ichthyomyine rodents (Muroidea): patterns of morphological evolution in a small adaptive radiation. – Bull. Amer. Mus. nat. Hist., New York, 188: 259–493.

***Vshivkova, T.S.*** (1979): (new data on the systematics and distribution of Megaloptera (Insecta) from the Soviet Far East). In: ***Levanidov, V.Ya.*** (ed.) Systematics and ecology of fish from continental waters of Soviet Far East, pp. 78–85. – Soviet Far East Scientific Centre, Vladivoslok.

– (1980): Alderflies (Megaloptera, Sialidae) of Mongolia and south Siberia. – Nasek. Mongol., Leningrad, 7: 283–288.

– (1985): Sialidae (Megaloptera) of Europe and the Caucasus. – Ent. Obozr., Moskva, 64: 146–157.

***Walcott, B. & Horridge, G.A.*** (1971): The compound eye of *Archichauliodes* (Megaloptera). – Proc. Roy. Soc. Lond. B., 179: 65–72.

***Walsh, B.D.*** (1863): Notes on Neuroptera. – Proc. ent. Soc. Phil., 15: 182–272.

***Walsh, B.D. & Riley, C.V.*** (1868): Notes. The hellgrammite fly. – Amer. Ent., St. Louis, Mo., 1: 61–62.

***Warner, R.W.*** (1971): Distribution of biota in a stream polluted by acid minedrainage. – Ohio J. Sci., 71: 202–216.

***Weber, H.*** (1928): Die Gliederung der Sternopleuralregion des Lepidopterenthorax. Eine vergleichend-morphologische Studie zur Subcoxaltheorie. – Z. Zool. Leipzig, 131: 181–254.

– (1933): Lehrbuch der Entomologie. Jena.

***Weed, C.M.*** (1889): Studies in pond life. Tech. Bull. Ohio Exp. Sta., 1: 4–17.

***Weidner, H.*** (1958): Einige interessante Insektenlarven aus der Bernsteininklusen-Sammlung des Geologischen Staatsinstituts Hamburg (Odonata, Coleoptera, Megaloptera, Planipennia). – Mitt. Geol. Staatsinst. Hamburg, 27: 50–68.

***Wickham, H.F.*** (1903): Meristic variation in *Corydalis cornuta*, Linn. – Can. Ent., Ottawa, 35: 207.

***Withycombe, C. L.*** (1925): Some aspects of the biology and morphology of the Neuroptera with special reference to immature stages and their possible phylogenetic significance. – Trans. ent. Soc. Lond. 1924: 303–412.

***Wood-Mason, J.*** (1884): Description of an Asian species of the neuropterous genus *Corydalis*. – Proc. Zool. Soc. Lond., 1884: 110.

***Woodrum, J. E. & Tarter, D. C.*** (1973): The life history of the alderfly, *Sialis aequalis* Banks, in an acid mine stream. – Amer. Midl. Nat., Notre Dame, Ind., 89: 360–368.

***Yamamoto, G.*** (1972): Trophic structure in Lake Tatsu-Nama, an acidotrophic lake in Japan, with special reference to the importance of the terrestrial community. In ***Kajak, Z. & Hillbricht-Ilkowska, A.*** (eds.). – Productivity Problems of Freshwaters, pp 405–419. IBP/UNESCO, Polish Science Publishers, Warsaw.

***Yamamoto, K. & Yoshiro, T.*** (1975): The fine structure of the lateral ocellus of the Dobsonfly larva. – J. Morph., Philadelphia, 146: 415–430.

***Yang, C-K & Yang, D.*** (1986): New fishflies from Guangxi, China. – Entomotaxonomia, Shensi, 8: 85–96.

– (1988): New species of Corydalinae from Yunnan (Megaloptera: Corydalinae). – Zool. Res., Beijing, 9: 45–60.

– (1989): The Dobsonflies of Shaanxi (Megaloptera: Corydalidae). – Entomotaxonomia, Shensi, 11: 9–11.

– (1990): The fishflies from Hainan Island (Megaloptera: Corydalidae). – Acta Zootax. Sin., Pei-ching, 15: 98–100.

***Yoshida, T.*** (1979): Life of a dobsonfly, *Protohermes grandis* – how the larva grows. – Insectarium, 16: 104–108.

***Yoshida, T., Sugimoto, K. & Hayashi, F.*** (1985): Notes on the life history of the Dobsonfly, *Protohermes grandis* Thunberg (Megaloptera, Corydalidae). – Kontyu, Tokyo, 53: 734–742.

***Zalessky, M. D.*** (1926): Observations sur un nouvel Insecte fossile du Permien de Kargala. – Bull. Soc. geol. Fr., Paris, 26: 75–84.

# Index of Scientific Names

Note: names listed are as in references cited in the text: especially for non-taxonomic work, a few errors may not have been detected. Numbers following a dash (–) page numbers; bold numbers are figure numbers

# General Index

**Hinweise für die Autoren**

**Handbuch der Zoologie – Handbook of Zoology**
**Band IV Arthropoda: Insecta**

Verlag: Walter de Gruyter, Berlin
Herausgeber: Maximilian Fischer, Wien

Das Handbuch der Zoologie, Band IV, Arthropoda: Insecta, behandelt das gesamte Gebiet der Entomologie. Es wird der gegenwärtige Stand der Kenntnisse in kurzer Form dargestellt, wobei vor allem die wesentlichen Tatsachen Berücksichtigung finden. Auf wichtige, noch zu klärende Fragen wird hingewiesen.

Band IV erscheint in Teilbänden mit fortlaufender Numerierung entsprechend dem Eingang der Beitragsmanuskripte.

Die Teilbände behandeln die Entomologie aus der Sicht verschiedener biologischer Teildisziplinen (Morphologie, Physiology, Tiergeographie, Phylogenie) oder von angewandten Gesichtspunkten (Medizin, Kultur, usw.) aus, sowie die Ordnungen des Insektenreiches.

Der Umfang der Manuskripte wird gemeinsam mit dem Herausgeber und den Mitarbeitern im Einverständnis mit dem Verlag festgelegt.

Jeder Teilband setzt sich zusammen aus:
- Inhaltsübersicht mit Seitenangabe
- Text
- Abbildungen (fortlaufend numeriert)
- Legenden zu den Abbildungen
- Tabellen (fortlaufend numeriert)
- Verzeichnis der wissenschaftlichen Namen (Gattungs- und Artnamen mit Autor und Jahreszahl, der Name des Autors entsprechend der Nomenklaturvorschrift – gegebenenfalls in Klammern)
- Sachregister (mit Seitenangaben), alphabetisch geordnet
- Literaturverzeichnis.

Text: Das Manuskript soll mit Schreibmaschine (2facher Zeilenabstand) auf DIN-A4-Blättern (21,0 cm × 29,7 cm) geschrieben werden. An der linken Seite soll ein 3 cm breiter Rand freigelassen werden.

Der Herausgeber bereitet das Manuskript für den Satz vor. Die Markierungen für die Auswahl der Schriftarten besorgt ebenfalls der Herausgeber. Die wissenschaftlichen Namen der Gattungen und Arten werden kursiv gesetzt. Sie werden mit Bleistift gewellt unterstrichen oder mit einem fluoreszierenden Farbstift markiert. Vorschläge für Kleindruck sollen durch einen senkrechten Bleistiftstrich am linken Rand mit der Anmerkung „petit" gemacht werden. Die Namen der Autoren der Taxa sollen im Text vermieden werden. Literatur-Hinweise erfolgen entweder durch Angabe der Nummer des fortlaufend numerierten Literaturverzeichnisses in [eckigen] Klammern oder Nennung des Autors mit Jahreszahl. Die Positionen der Abbildungen sollen am Rand des Manuskriptes mit Bleistift eingetragen werden.

Die wissenschaftlichen Periodica sollen im Literaturverzeichnis nach der „World List of Scientific Periodicals, published in the Years 1900–1960 (Butterworths, London)" oder nach den nachfolgenden Listen zitiert werden.

Abbildungsvorlagen (Zeichnungen und Fotos) müssen in reproduktionsfähigem Zustand geliefert werden.

Die Manuskripte sollen in 3 Teilen vorbereitet und eingereicht werden:
1. vollständiger Text
2. alle Abbildungen, Tabellen und Diagramme
3. sämtliche Legenden zu den Abbildungen.

Der Autor soll eine Kopie des Manuskriptes anfertigen und bei sich behalten.

Als Sprachen sind Deutsch und Englisch zugelassen.

Die Manuskripte, Fahnen und der Umbruch sind an den Herausgeber zu senden.

*Herausgeber und Schriftleiter*

Hofrat Univ.-Doz.
Mag. Dr. Maximilian Fischer
Direktor am
Naturhistorischen Museum Wien
Burgring 7
A-1014 Wien
Austria
Telefon (0222) 934541-316

*Verlag*

Walter de Gruyter
Genthiner Straße 13
D-1000 Berlin (West) 30
F. R. of Germany
Tel. (030) 26005-124

Walter de Gruyter, Inc.
Scientific Publishers
200 Saw Mill River Road
Hawthorne, N. Y. 10532
USA Tel. (914) 747-0110

## Handbuch der Zoologie, Band IV Arthropoda: Insecta

Band IV Arthropoda: Insecta wird von nun an in Teilbänden mit einer fortlaufenden Numerierung erscheinen, die an die bisherige Zählung der Lieferungen anschließt. Jeder Teilband wird von 28 an gebunden geliefert. Für die bereits erschienenen 27 Lieferungen wird zum Binden nachfolgender Plan empfohlen.

### Bisher sind erschienen

| Autor | Titel | Beitrag | Lieferung |
|---|---|---|---|
| Alfred Kaltenbach | Embiodea (Spinnfüßer) | 8 | 1 |
| Fritz Zumpt | Human- und veterinärmedizinische Entomologie | 9 | |
| Erwin Schimitschek | Insekten als Nahrung, in Brauchtum, Kult und Kultur | 10 | 2 |
| Douglas St. Quentin & Max Beier | Odonata (Libellen) | 6 | 3 |
| Max Beier | Mantodea (Fangheuschrecken) | 12 | 4 |
| Hermann Priesner | Thysanoptera (Physopoda, Blasenfüßer) | 19 | 5 |
| Max Beier | Phasmida (Stab- oder Gespenstheuschrecken) | 10 | 6 |
| Joachim Illies | Ephemeroptera (Eintagsfliegen) | 5 | 7 |
| Maximilian Fischer | Die Verwandlung der Insekten | 16 | 8 |
| Max Beier | Diagnose | 1 | |
| Max Beier | Geschichte der Entomologie | 2 | |
| Boris B. Rohdendorf | Paläontologie | 3 | 9 |
| Boris B. Rohdendorf | Phylogenie | 4 | |
| Max Beier | Klassifikation | 5 | |
| Stefan von Kéler † | Mallophaga (Federlinge u. Haarlinge) | 17 | 10 |
| Herbert Franz & Max Beier | Die geographische Verbreitung der Insekten | 6 | 11 |
| Friedrich Schaller | Collembola (Springschwänze) | 1 | 12 |
| Herbert Weidner | Isoptera (Termiten) | 14 | |
| Herbert Weidner | Zoraptera (Bodenläuse) | 15 | 13 |
| Heinz Janetschek | Protura (Beintastler) | 3 | 14 |
| Ragnar Kinzelbach | Strepsiptera (Fächerflügler) | 24 | |
| Horst Aspöck & Ulrike Aspöck | Raphidioptera (Kamelhalsfliegen) | 25 | 15 |
| Karl H. C. Jordan | Heteroptera (Wanzen) | 20 | 16 |
| Max Beier | Saltatoria (Heuschrecken u. Grillen) | 9 | 17 |
| Herbert Weidner | Copeognatha (Staubläuse) | 16 | 18 |
| Erwin Schimitschek | Pflanzen-, Vorrats- und Materialschädlinge | 8 | 19 |
| Willi Hennig | Diptera (Zweiflügler) | 31 | 20 |
| Hans Malicky | Trichoptera (Köcherfliegen) | 29 | 21 |
| Max Beier | Blattariae (Schaben) | 13 | 22 |
| Klaus Günther & Konrad Herter | Dermaptera (Ohrwürmer) | 11 | 23 |
| Felix Mainx | Zytologie (Karyologie) | 12 | 24 |
| Alfred Kaltenbach | Mecoptera (Panorpata, Schnabelhafte) | 28 | 25 |
| Peter Zwick | Plecoptera (Perlaria, Uferfliegen) | 7 | 26 |
| Herbert Weidner | Morphologie, Anatomie und Histologie | 11 | 27 |
| Hans Strümpel | Homoptera (Pflanzensauger) | Teilband 28 | |
| Werner Nachtigall | Bewegungsphysiologie (Laufen·Schwimmen·Fliegen) | Teilband 29 | |
| Timothy R. New | Planipennia (Netzflügler) | Teilband 30 | |
| Wolfgang Schedl | Hymenoptera: Unterordnung Symphta (Pflanzenwespen) | Teilband 31 | |
| Feliks Piotrowski | Anoplura (echte Läuse) | Teilband 32 | |
| Timothy R. New & Günther Theischinger | Megaloptera (Alderflies, Dobsonflies) | Teilband 33 | |

**Bindevorschlag für das Handbuch der Zoologie**
**Band IV Arthropoda: Insecta**

| Beitrag Nr. | | Lieferung Nr. |
|---|---|---|
| | Beiträge auf die Entwicklung bezogen | |
| 1 | M. Beier: Diagnose | 9 |
| 2 | M. Beier: Geschichte der Entomologie | 9 |
| 3 | B. B. Rohdendorf: Paläontologie | 9 |
| 4 | B. B. Rohdendorf: Phylogenie | 9 |
| 5 | M. Beier: Klassifikation | 9 |
| 6 | H. Franz & M. Beier: Die geographische Verbreitung der Insekten | 11 |
| 16 | M. Fischer: Die Verwandlung der Insekten | 8 |
| | (zusammen 291 Seiten) | |
| | Beitrag auf die Form bezogen | |
| 11 | H. Weidner: Morphologie, Anatomie, Histologie | 27 |
| | (531 Seiten) | |
| | Beitrag auf „Form" im weiteren Sinn bezogen | |
| 12 | F. Mainx: Zytologie (Karyologie) | 24 |
| | (130 Seiten) | |
| | Auf den Menschen bezogen | |
| 8 | E. Schimitschek: Pflanzen-, Material- und Vorratsschädlinge | 19 |
| 9 | F. Zumpt: Human- and veterinärmedizinische Entomologie | 2 |
| 10 | E. Schimitschek: Insekten als Nahrung, in Brauchtum, Kult und Kultur | 2 |
| | (zusammen 311 Seiten) | |
| | Apterygota und ursprüngliche Pterygota | |
| 3 | H. Janetschek: Protura (Beintastler) | 14 |
| 1 | F. Schaller: Collembola (Springschwänze) | 12 |
| 5 | J. Illies: Ephemeroptera (Eintagsfliegen) | 7 |
| 6 | D. St. Quentin & M. Beier: Odonata (Libellen) | 3 |
| | (zusammen 246 Seiten) | |
| | Einzelne Ordnung | |
| 7 | P. Zwick: Plecoptera (Steinfliegen) | 26 |
| | (111 Seiten) | |
| | Isolierte Ordnungen der Paurometabola | |
| 8 | A. Kaltenbach: Embiodea (Spinnfüßer) | 1 |
| 11 | K. Günther & K. Herter: Dermaptera (Ohrwürmer) | 23 |
| | (zusammen 187 Seiten) | |
| | Blattopteroidea = Dictyoptera | |
| 12 | M. Beier: Mantodea (Fangheuschrecken) | 4 |
| 13 | M. Beier: Blattariae (Schaben) | 22 |
| 14 | H. Weidner: Isoptera (Termiten) | 13 |
| | (zusammen 321 Seiten) | |
| | Orthopteroidea | |
| 10 | M. Beier: Phasmida (Stab- oder Gespenstheuschrecken) | 6 |
| 9 | M. Beier: Saltatoria (Heuschrecken und Grillen) | 17 |
| | (zusammen 217 Seiten) | |
| | Bisher erschienene Paraneoptera, ausgenommen Homoptera | |
| 15 | H. Weidner: Zoraptera (Bodenläuse) | 13 |
| 16 | H. Weidner: Copeognatha (Staubläuse) | 18 |
| 19 | H. Priesner: Thysanoptera (Physopoda, Blasenfüßer) | 5 |
| 20 | K. Jordan: Heteroptera (Wanzen) | 16 |
| 17 | St. v. Kéler: Mallophaga (Federlinge und Haarlinge) | 10 |
| | (zusammen 323 Seiten) | |

| Beitrag Nr. | | Lieferung Nr. |
|---|---|---|
| | Einzelne große Ordnung | |
| 31 | W. Hennig: Diptera (Zweiflügler) | 20 |
| | (337 Seiten) | |
| | Restliche bisher erschienene Holometabola | |
| 25 | H. & U. Aspöck: Raphidioptera (Kamelhalsfliegen) | 15 |
| 29 | H. Malicky: Trichoptera (Köcherfliegen) | 21 |
| 28 | A. Kaltenbach: Mecoptera (Panorpata, Schnabelhafte) | 25 |
| 24 | R. Kinzelbach: Strepsiptera (Fächerflügler) | 15 |
| | (zusammen 341 Seiten) | |

Bei Fragen zur Produktsicherheit wenden Sie sich bitte an:
If you have any questions regarding product safety,
please contact:

Walter de Gruyter GmbH
Genthiner Straße 13
10785 Berlin
productsafety@degruyterbrill.com